AF845230

Intelligent Data-Driven Techniques for Security of Digital Assets

The book covers the role of emerging technologies such as blockchain technology, machine learning, IoT, cryptography, etc., in digital asset management. It further discusses the digital asset management applications in different domains such as healthcare, travel industry, image processing, and our daily life activities to maintain privacy and confidentiality.

This book:

- Discusses techniques for securing and protecting digital assets in collaborative environments, where multiple organizations need access to the same resources.
- Explores how artificial intelligence can be used to automate the management of digital assets, and how it can be used to improve security and privacy.
- Explains the role of emerging technology such as blockchain technology for transforming conventional business models.
- Highlights the importance of machine learning technique in maintaining the privacy and security of data.
- Covers encryption and decryption techniques, their advantages and role in improving the privacy of data.

The text is primarily written for senior undergraduates, graduate students, and academic researchers in diverse fields including electrical engineering, electronics and communications engineering, computer science and engineering, information technology, and business management.

Intelligent Data-Driven Systems and Artificial Intelligence
Series Editor: Harish Garg

Cognitive Machine Intelligence
Applications, Challenges, and Related Technologies
Inam Ullah Khan, Salma El Hajjami, Mariya Ouaissa, Salwa Belqziz and Tarandeep Kaur Bhatia

Artificial Intelligence and Internet of Things based Augmented Trends for Data-Driven Systems
Anshu Singla, Sarvesh Tanwar and Pao-Ann Hsiung

Modelling of Virtual Worlds Using the Internet of Things
Edited by Simar Preet Singh and Arun Solanki

Data-Driven Technologies and Artificial Intelligence in Supply Chain
Tools and Techniques
Mahesh Chand, Vineet Jain and Puneeta Ajmera

Intelligent Data-Driven Techniques for Security of Digital Assets
Edited by Arun Kumar Rana, Sumit Kumar Rana, Vishnu Sharma, and Ritu Dewan

For more information about this series, please visit: www.routledge.com/Intelligent-Data-Driven-Systems-and-Artificial-Intelligence/book-series/CRCIDDSAAI

Intelligent Data-Driven Techniques for Security of Digital Assets

Edited by
Arun Kumar Rana
Sumit Kumar Rana
Vishnu Sharma
Ritu Dewan

CRC Press
Taylor & Francis Group
Boca Raton London New York

CRC Press is an imprint of the
Taylor & Francis Group, an **informa** business

Designed cover image: shutterstock

First edition published 2025
by CRC Press
2385 NW Executive Center Drive, Suite 320, Boca Raton FL 33431

and by CRC Press
4 Park Square, Milton Park, Abingdon, Oxon, OX14 4RN

CRC Press is an imprint of Taylor & Francis Group, LLC

© 2025 selection and editorial matter, Arun Kumar Rana, Sumit Kumar Rana, Vishnu Sharma and Ritu Dewan; individual chapters, the contributors

Reasonable efforts have been made to publish reliable data and information, but the author and publisher cannot assume responsibility for the validity of all materials or the consequences of their use. The authors and publishers have attempted to trace the copyright holders of all material reproduced in this publication and apologize to copyright holders if permission to publish in this form has not been obtained. If any copyright material has not been acknowledged please write and let us know so we may rectify in any future reprint.

Except as permitted under U.S. Copyright Law, no part of this book may be reprinted, reproduced, transmitted, or utilized in any form by any electronic, mechanical, or other means, now known or hereafter invented, including photocopying, microfilming, and recording, or in any information storage or retrieval system, without written permission from the publishers.

For permission to photocopy or use material electronically from this work, access www.copyright.com or contact the Copyright Clearance Center, Inc. (CCC), 222 Rosewood Drive, Danvers, MA 01923, 978-750-8400. For works that are not available on CCC please contact mpkbookspermissions@tandf.co.uk

Trademark notice: Product or corporate names may be trademarks or registered trademarks and are used only for identification and explanation without intent to infringe.

ISBN: 978-1-032-74488-9 (hbk)
ISBN: 978-1-032-96685-4 (pbk)
ISBN: 978-1-003-59043-9 (ebk)

DOI: 10.1201/9781003590439

Typeset in Sabon
by SPi Technologies India Pvt Ltd (Straive)

Contents

About the Editors

Arun Kumar Rana completed his B.Tech degree from Kurukshetra University, M.Tech. and Ph.D. degrees from Maharishi Markandeshwar (Deemed to be University), Mullana, India. His areas of interest include Image Processing, Wireless Sensor Network, Internet of Things, AI, and Machine Learning and Embedded systems. Prof. Rana is currently Associate Professor at Galgotias College of Engineering and Technology, Greater Noida, India with more than 18 years of experience. Prof. Rana is a collaborative researcher He has published 35 SCI Papers (6 submitted), 30 SCOPUS papers, 25 chapters, 30 Papers in National and International conferences. He has also published 10 books with national and international publishers and has been a member of many SCI, Scopus indexed international conferences/symposiums.

Sumit Kumar Rana completed his B.Tech degree from Kurukshetra University, M.Tech. and Ph.D. degree from Maharishi Markandeshwar (Deemed to be University), Mullana, India. His areas of interest include Blockchain technology, cryptography, cryptocurrency, artificial intelligence etc. Prof. Sumit Kumar Rana is currently an Associate Professor at University Institute of Engineering Chandigarh University, Punjab, India with more than 13 years of experience. Prof. Rana is a collaborative researcher. He has published multiple SCI/SCOPUS papers, book chapters and papers in National and International IEEE conferences. He is also editing multiple books with reputed international publishers.

Vishnu Sharma is Professor and Dean of CSE at ITS College of Engineering, Greater Noida, India. Dr. Sharma completed his BTech, MTech, and PhD (CSE) in 2012 from the Government Autonomous Institute, Madhav Institute of Technology and Science (M.I.T.S.), Gwalior, MP, in Computer Science and Engineering, and was affiliated with the Rajiv Gandhi Technical University, Bhopal, India. His areas of interest are mobile computing, cybersecurity, and advanced mobile computing.

Ritu Dewan earned her B.Tech. from Kurukshetra University and M.Tech. from Thapar University. She is currently pursuing her PhD (Computer Science) from Lingyas Vidyapeeth. She has registered 10 designs and published more than 25 patents. She has published various research papers in reputed international journals and conferences. Her areas of expertise are Broadcast networks, IoT, AI and Blockchain. Prof. Ritu is currently an Assistant Professor at Galgotias college of engineering and technology, Greater Noida, India with more than 10 years of experience. She is also Guest editor for Special Issue ECSI, Scopus index journal.

Contributors

Gaurav Agrawal
Galgotias University
Greater Noida, India

Ashok
Government P.G College
Ambala Cantt, India

Ravi Kumar Barwal
Government College for Women
Shahzadpur, India

Sana Bharti
Chitkara University Institute of Engineering and Technology
Chitkara University
Punjab, India

Anupam Bonkra
Maharishi Markandeshwar (Deemed to be University)
Ambala, India

G. Boopathi Raja
Velalar College of Engineering and Technology
Erode, India

Ajay Beniwal
University of Glasgow
U.K.

Anjali Chauhan
KIET Group of Institutions
Delhi-NCR, Ghaziabad, India

Ritu Dewan
Galgotias College of Engineering and Technology
Greater Noida, India

Pummy Dhiman
Chitkara University Institute of Engineering and Technology
Chitkara University
Punjab, India

Anurag Gupta
KIET Group of Institutions
Delhi-NCR, Ghaziabad, India

Kavya Gupta
KIET Group of Institutions
Delhi-NCR, Ghaziabad, India

Pooja Gupta
Galgotias College of Engineering and Technology
Greater Noida, India

Chhavi Jain
Sharda University
Greater Noida, India

Jaya
Panipat Institute of Engineering and Technology
Panipat, Haryana, India

Roshni Jeewani
Galgotias College of Engineering and Technology
Greater Noida, India

Ramandeep Kaur
CT Institute of Technology and Research
Jalandhar, India

Sunny Kuhar
Panipat Institute of Engineering and Technology
Panipat, India

Naveen Kumar
National Institute of Technology
Kurukshetra, India

Kamal
Kurukshetra University
Kurukshetra, Haryana, India

Amit Kumar Kesarwani
Galgotias College of Engineering and Technology
Greater Noida, India

Jhansi Bharathi Madavarapu
University of the Cumberlands
Williamsburg, KY

Manisha
Panipat Institute of Engineering and Technology
Panipat, Haryana, India

Abha Kiran Rajpoot
KIET Group of Institutions
Delhi-NCR, Ghaziabad, India

Komal Rani
Babu Anant Ram Janta College
Kaithal, Haryana, India

Manju Rani
Gurugram University
Gurugram, India

Arun Rana
Galgotias College of Engineering and Technology
Greater Noida, India

Dr Sumit Kumar Rana
Associate Professor
Department of Computer Science & Engineering
University Institute of Engineering
Chandigarh University, Mohali, Punjab, India

Neeraj Rohilla
Government College
Barwala, India

Sarah
Galgotias College of Engineering and Technology
Greater Noida, India

Wasswa Shafik
University Brunei Darussalam
Bandar Seri Begawan, Brunei Darussalam

Bhavna Sharma
Chitkara University Institute of Engineering and Technology
Chitkara University
Punjab, India

Inderpal Singh
CT Institute of Technology and Research
Jalandhar, India

Shikha Singh
Galgotias College of Engineering and Technology
Greater Noida, India

Jai Sonker
Amity School of Hospitality
Amity University
Rajasthan, India

Kuldeep Vayadande
Vishwakarma Institute of Technology
Pune, Maharashtra, India

Navneet Verma
Panipat Institute of Engineering and Technology
Panipat, India

Pawan Whig
Vivekananda Institute of Professional Studies
New Delhi, India

Balaram Yadav Kasula
University of the Cumberlands
Williamsburg, KY

Nikhitha Yathiraju
University of the Cumberlands
Williamsburg, KY

Chapter 1

Integrating emerging technologies for digital asset management

Wasswa Shafik

1.1 INTRODUCTION

Digital asset management (DAM) plays a decisive role in global digital economies, functioning as a catalyst for innovation, asset employment generation, and overall economic advancement [1]. These firms encompass a wide range of industries and are dynamic since they are significant in supplying critical goods and services to local and worldwide markets. Recently, there has been a significant shift in the corporate environment, primarily influenced by the rapid progression of technology [2]. Technology-enabled business procedures are essential in fostering business innovation and economic growth within the contemporary business financial transaction environment. DAM encounters several obstacles in the current highly competitive business landscape. Organizations face the challenges of managing limited resources, restricted capital availability, and the imperative of emerging technology adaptability in response to dynamic market requirements [3].

Artificial intelligence (AI)-enabled chatbot technology, for example, ChatGPT, presents a notable opportunity for DAM to augment customer engagement, optimize communication processes, and deliver tailored experiences [4]. Blockchain technology, renowned for its decentralized and tamper-resistant ledger system, holds the potential to enhance transparency, security, and trust within the operations of DAM. This technology presents opportunities for fostering collaboration and improving operational efficiency. Using robotics can enable This, in turn, to lead to enhanced delivery of products and services [5]. These technologies are propelled by data analytics and machine learning, provide DAM with the means to acquire practical knowledge, streamline decision-making procedures, and maintain competitiveness in a progressively reliant marketplace.

ChatGPT, a prominent instance of natural language processing (NLP), can potentially significantly transform customer interactions and knowledge management within DAM [6]. The ability of this technology to offer prompt and human-like responses to inquiries has an impact on enlightening customer service, optimizing operational processes, and promoting active participation [7]. While exploring the possibilities of this development, it is imperative

DOI: 10.1201/9781003590439-1

to carefully manage the ethical implications of data privacy protection and content accuracy. Blockchain, the foundational technology of cryptocurrencies, presents DAM with a novel prospect to augment transparency, security, and confidence in transactions and managing supply chains [8]. Using a decentralized ledger system can substantially mitigate fraud and errors, enhancing the efficacy of cross-border transactions. However, using blockchain technology in DAM presents complex obstacles, such as issues related to scalability, adherence to regulatory requirements, and gaps in expertise.

Robotics, distinguished by its ability to automate tasks and perform physical actions, offers a compelling opportunity for DAM to achieve cost-effective process optimization and reduce personnel requirements [8]. Figure 1.1 presents eight essential emerging technologies, including robots that execute repeated activities accurately, enhance production, and diminish operational expenditures. It is vital for DAM to thoroughly evaluate the initial capital expenditure needed to cater to the business's sets and consider providing additional training to their employees to integrate these automated systems effectively [6]. AI provides valuable insights based on data analysis, automates various processes, and brings innovative advancements to DAM [5].

The digital age has initiated a novel epoch in corporate operations, wherein technological advancements have driven market dynamics and client demands to unparalleled levels. Within this context, DAM faces the challenge of operating within a multifaceted environment [1]. Not only do they encounter competition from well-established corporations, but they also endeavor to give a competitive edge through the timely adoption of technical advancements. The rapidity of change can be disorienting, and the consequences are significant [2]. DAM is at a critical juncture characterized by favorable prospects and formidable obstacles. On the one hand, they demonstrate the ability to quickly adapt to changes in the market and adjust their strategies according to evolving consumer demands [3]. However,

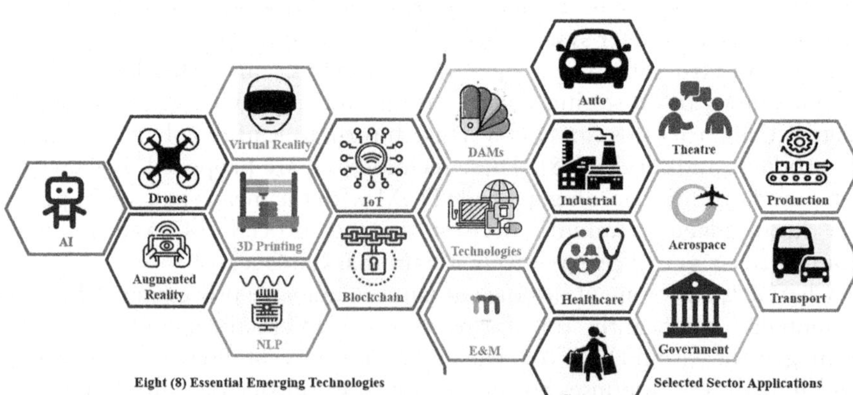

Figure 1.1 Emerging technologies and their application in different sectors.

limited marketing resources, constrained budgets, and a potentially challenging learning curve might present substantial barriers to general technological innovation and digital transformation [4].

1.1.1 The chapter organization

The chapter contributions are listed below.

- The study illustrates ChatGPT overview, application, and how it can benefit DAM, demonstrated by provided case studies coupled with a discussion on the current challenges of its adoption.
- Presents Blockchain for DAM, describing the fundamentals of Blockchain technology and exploring how it can enhance security, transparency, and efficiency for DAM. It also presents real-world applications of Blockchain in DAM settings and describes potential barriers to Blockchain adoption.
- Details the robotics for DAM, presenting the robotics and automation technologies, discussing how robotics can streamline processes and reduce costs for DAM, and providing instances of DAM implementing robotics solutions and DAM considerations.
- Explores these technology integrations for more significant impact, discussing the potential amalgamation using existing cases and providing insights into how DAM can create a holistic technology strategy.
- Finally, it presents the challenges, limitations, and considerations, summarizing common challenges DAM may face when adopting these technologies and the emerging trends in technology adoption for DAM, predicting the future landscape of DAM with these technologies.

1.1.2 The chapter organization

The remainder of this chapter is structured as follows. Section 1.2 presents ChatGPT for DAM, explaining what ChatGPT is and how it can benefit DAM, providing case studies of DAM successfully using ChatGPT, and discussing the potential challenges and limitations of ChatGPT adoption by DAM. Section 1.3 demonstrates Blockchain for DAM, detailing the fundamentals of blockchain technology, explores how Blockchain can enhance transparency, security, and efficiency for DAM, real-world applications of Blockchain in DAM settings, and discusses potential barriers to blockchain adoption by DAM. Section 1.4 details robotics for DAM, introducing robotics and automation technologies, discussing how robotics can streamline processes and reduce costs for DAM, providing instances of DAM implementing robotics solutions, and addressing considerations, limitations, and challenges for DAM looking to embrace robotics. Section 1.5 demonstrates the synergy and integration and explores how these technologies can be integrated for a more significant impact, discussing potential integration and complementary use cases and providing insights into how DAM can create a holistic technology strategy. Section 1.6 presents the considerations and challenges, summarizing common challenges DAM may face when adopting these technologies. Finally, Section 1.7. presents the emerging trends in the adoption of technology for DAM, predicting the future landscape of DAM with these technologies.

1.2 ChatGPT FOR DAM

ChatGPT is an advanced AI model that OpenAI has developed, and it belongs to the broader category of models referred to as Generative Pre-trained Transformers (GPT). ChatGPT has been purposefully developed to excel in comprehending and producing natural language, enabling it to digest and generate text that resembles human-like communication effectively, contingent upon the input it is provided with [9]. The software possesses numerous essential attributes that contribute to its efficacy as a tool for comprehending and producing writing closely resembling human language. The subsequent points outline the principal characteristics of ChatGPT.

1.2.1 Key features of ChatGPT

The comprehensive attributes of ChatGPT enable it to perform exceptionally well in a wide range of conversational AI applications, encompassing customer assistance, question answering, content generation, and users in intricate activities. The valuable attributes of adaptability, scalability, and context-awareness make it a highly sought-after resource for enterprises to utilize advanced natural language interpretation and generation capabilities.

ChatGPT has been specifically developed to engage in discussions that emulate real human interactions, enabling it to respond to user inquiries and messages [10]. Comprehending and producing text within a conversational framework is paramount for various applications such as chatbots, virtual assistants, and customer care systems. An exemplary attribute of ChatGPT is its adeptness at comprehending and upholding contextual coherence throughout a dialogue [11]. The language model can recall and cite preceding messages, generating contextually appropriate and coherent responses within the continuing discourse.

The language generated by ChatGPT exhibits coherence and contextual relevance, rendering it well-suited for many activities, including composing emails, providing answers, and participating in conversations. The system can generate replies that exhibit a natural and human-like quality. ChatGPT can manage multi-turn discussions, enabling it to participate actively in iterative exchanges with users [12]. Users inquire about further information, shift the subject matter, or offer supplementary details, and ChatGPT will answer accordingly. Customization refers to modifying or tailoring a product or

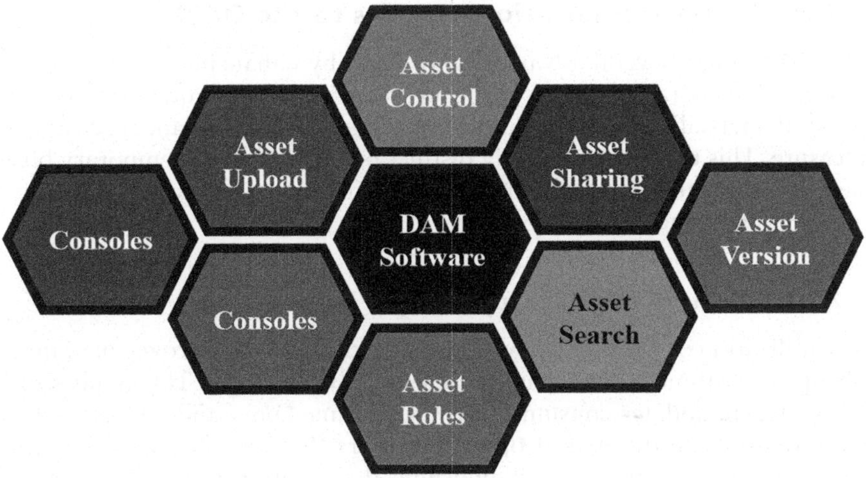

Figure 1.2 DAM software usage.

service according to individual preferences or specific requirements [12]. Although ChatGPT is initially trained on a comprehensive dataset, it can be customized to specific applications or sectors. The customization feature enables enterprises to tailor ChatGPT to their specific requirements, encompassing domain-specific language and rules [11].

Prompt flexibility refers to the ability of a prompt or question to accommodate various interpretations and responses, as some DAM software can be of use, as illustrated in Figure 1.2. It allows individuals to approach the prompt from different angles and provide diverse perspectives or solutions; users can engage ChatGPT by submitting prompts or messages, which may exhibit a range of lengths and levels [12]. The system's inherent flexibility allows users to commence conversations and tasks in a manner that aligns with their requirements. Large-scale knowledge is a comprehensive and extensive understanding of a specific subject or field. It involves accumulating and integrating a significant amount of information and data. ChatGPT has undergone pre-training using an extensive corpus of textual data from the internet, equipping it with a vast repository of general knowledge. The system can furnish information and respond to inquiries about various subjects [9].

Although ChatGPT is primarily designed for text-based inputs, it can process inputs that include additional context, such as photos or documents. Incorporating multi-modal capabilities enables users to offer a more extensive range of information, facilitating more comprehensive interactions [13]. ChatGPT has been specifically developed to effectively manage and accommodate a substantial quantity of concurrent chats, rendering it well-suited for enterprises or platforms with significant user engagement levels. The system's scalability ensures its capacity to accommodate the increasing demands of expanding user populations.

1.2.2 ChatGPT application benefits to the DAM

ChatGPT offers several advantages to DAM by enhancing customer assistance, enhancing operational efficiency, reducing costs, acquiring vital customer data, facilitating personalization, and, ultimately, enhancing customer pleasure. This tool grants DAM a distinct benefit in the contemporary business milieu, enabling it to prosper and expand amidst a fiercely competitive and ever-changing setting.

1.2.2.1 Enhanced customer support

The utilization of ChatGPT-powered chatbots enables the provision of uninterrupted customer service throughout the day and night. This implies that DAM accommodates consumers in various time zones and addresses their concerns outside the typical operating hours [9]. ChatGPT can promptly reply to consumer inquiries, diminishing the duration of response periods and guaranteeing that customers obtain promptly at critical moments. ChatGPT can effectively alleviate the burden on human customer support workers by managing regular questions and duties, reducing workload [10]. This enables employees to direct their attention towards more intricate matters and assignments that necessitate human experience. The entire five DAM lifecycle is presented in Figure 1.3.

1.2.2.2 Scalability

ChatGPT can be advantageous for DAM since it enables them to handle an escalating volume of consumer contacts effectively without necessitating the recruitment of supplementary personnel. The scalability of this system offers both cost-effectiveness and flexibility [13]. Consistency is a crucial feature of ChatGPT since it ensures that client inquiries are met with consistently

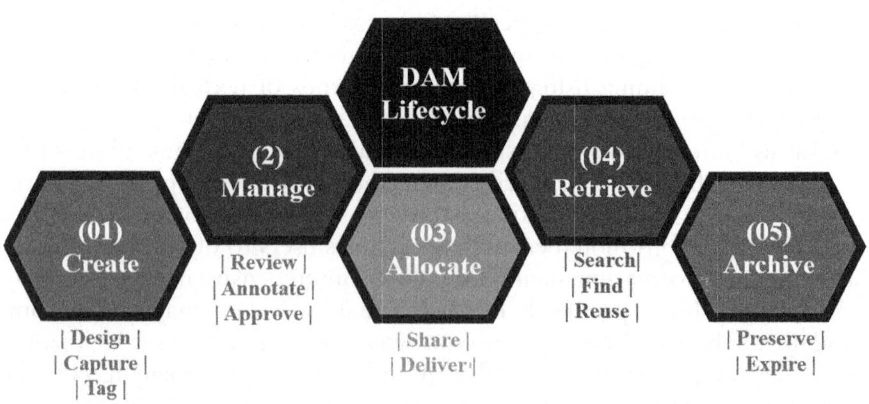

Figure 1.3 DAM lifecycle.

reliable and uniform responses. All customers are provided with an equal service and information standard, minimizing the potential for human errors and inconsistencies.

1.2.2.3 Cost savings

The potential reduction of labor costs for DAM can be achieved through automation in customer service and regular operations, hence eliminating the need to hire and train more staff [14]. Operational efficiency is enhanced with the implementation of ChatGPT, which effectively streamlines various business operations. This has the potential to result in long-term cost reductions.

1.2.2.4 Data collection and analysis

ChatGPT gathers substantial data during this exchange, encompassing client preferences, prevalent difficulties, and comments. DAM now evaluates this data to obtain valuable insights about client behavior, enabling them to customize their products and services accordingly. ChatGPT can facilitate market research by collecting feedback on novel products or services, identifying emerging patterns, and assessing client attitudes.

1.2.2.5 Personalization

By implementing customization techniques, DAM utilizes ChatGPT to comprehend industry-specific jargon, enabling it to deliver customized responses to consumers. This phenomenon improves user satisfaction and establishes robust customer connections [11]. ChatGPT possesses the capability to evaluate client data in order to provide tailored suggestions for products or services, hence enhancing the potential for cross-selling and upselling.

1.2.2.6 Efficient handling of routine tasks

ChatGPT can assist clients in monitoring the progress of their orders by offering timely updates regarding the delivery status and estimated arrival time. Frequently asked questions (FAQs) can be effectively addressed by utilizing ChatGPT, allowing customer support professionals to allocate resources to resolve more intricate concerns. DAM operating in service industries can effectively utilize ChatGPT to facilitate appointment scheduling processes [15]. This technology enables clients to book desired services, conveniently enhancing the overall customer experience.

1.2.2.7 Improved customer satisfaction

The prompt highlights that providing prompt and precise responses to client inquiries is associated with elevated levels of customer satisfaction. Satisfied clients are more inclined to develop loyalty towards and advocate for the DAM [12]. Consistent service is crucial for cultivating trust and dependability, which are fundamental for establishing a favorable brand image.

1.2.2.8 Competitive advantage

DAM may achieve differentiation by utilizing ChatGPT for customer assistance and engagement. This enables them to stand out from their competitors by providing unique and responsive services [13]. The ability to adapt to prevailing trends, such as the utilization of ChatGPT, displays DAM's dedication to fulfilling consumer demands within the context of the digital era.

1.3 BLOCKCHAIN FOR THE DAM

Blockchain technology functions through a distributed network of computer nodes instead of being dependent on a singular central authority. Every node inside the network possesses a complete replica of the whole Blockchain, guaranteeing transparency and obviating the necessity for intermediaries [16]. Sophisticated cryptographic methods are employed by blockchain technology to ensure the security of data and transactions. Every transaction undergoes encryption and is connected to the preceding transaction using a sophisticated mathematical procedure known as hashing [17]. The cryptographic measures employed in this context guarantee the confidentiality and integrity of data stored on the Blockchain.

Once data is recorded on a blockchain, it undergoes a significant level of immutability, making it highly challenging to modify or erase. Every block within the system is equipped with the preceding block, establishing a sequential and interconnected chain of blocks [18]. The alteration of data within a single block necessitates the modification of all following blocks in the chain, a computationally impractical task owing to the network's decentralized structure. Consensus techniques are employed to authenticate and incorporate novel transactions into the Blockchain [19]. The prevailing mechanism utilized in many systems is known as Proof of Work (PoW), wherein miners compete to solve intricate mathematical puzzles. After successfully solving the computational puzzle, the proposed block is appended to the Blockchain, and miners receive their respective rewards. Proof of Stake (PoS) is an alternative consensus process in which validators are selected to generate new blocks based on the quantity of cryptocurrency they own and are prepared to "stake" as collateral [20].

Blockchain transactions exhibit a high degree of transparency, allowing them to be accessed and observed by any participant within the network. The verification of transaction validity by each participant serves to foster trust and accountability. Smart contracts can execute themselves based on

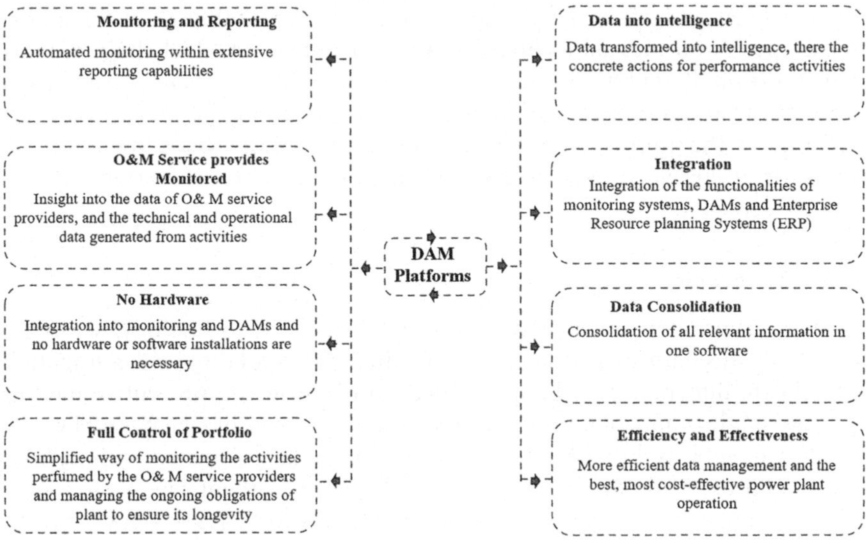

Figure 1.4 Centralized DAM platform.

predetermined criteria and conditions [21]. They are executed automatically when specific circumstances are satisfied without requiring intermediaries. Ethereum is a prominent blockchain platform renowned for its robust facilitation of smart contracts. In a blockchain network, individuals utilize digital signatures to authenticate their identity and verify the legitimacy of transactions [22]. Every individual involved in the process possesses a unique cryptographic private key, which is utilized to sign transactions. Also, they own a corresponding public key, which verifies the integrity and authenticity of those transactions; the centralized DAM platform is demonstrated in Figure 1.4.

1.3.1 DAM-enabled blockchain enhances transparency, security, and efficiency

Utilizing blockchain technology empowers DAM by augmenting transparency by establishing immutable records, strengthening security through encryption and decentralized architecture, and increasing efficiency through expedited transactions and automated operations. Adopting blockchain technology can yield cost savings, enhance process optimization, and foster heightened confidence among consumers and business partners.

1.3.1.1 Transparency

Utilizing blockchain technology facilitates enhanced transparency within the supply chain, enabling DAM to document each stage of production and distribution meticulously. Its implementation mitigates the prevalence of counterfeit goods [23]. Furthermore, it offers immutable records, guaranteeing precision and data integrity. The utilization of blockchain technology additionally facilitates the creation of an audit trail, hence streamlining financial transactions and enhancing compliance reporting. Smart contracts streamline the implementation of contractual obligations, removing the need for go-betweens and promoting greater transparency within commercial agreements.

1.3.1.2 Security

Data encryption plays a pivotal role in blockchain technology, serving as a critical mechanism for safeguarding the confidentiality and authenticity of delicate information. The use of decentralization is a preventive measure against single points of failure. Consensus mechanisms like PoW and Proof of Stake are employed to address and reduce the risks associated with network manipulation [24]. Private keys are utilized to authenticate transactions, while cryptographic keys offer robust procedures for transaction verification and user authentication. Utilizing blockchain and decentralized ledger also facilitates data recovery for DAM in the event of data loss or corruption.

1.3.1.3 Efficiency

Blockchain technology offers significant benefits for DAM in cross-border transactions. It accelerates payment processing, reduces settlement periods, and improves cash flow management. Blockchain technology eliminates intermediaries, reducing transaction costs and settlement time. Smart contracts automate agreements, reducing human involvement [17]. Blockchain technology also facilitates efficient record-keeping, reducing administrative burdens and errors. Blockchain-enhanced crowdfunding and peer-to-peer lending platforms provide DAM with direct financing, eliminating the need for conventional financial intermediaries [18].

1.4 ROBOTICS FOR DAM

Medium-sized enterprises are transforming their operational landscape by implementing automation technology, enabling increased production, reduced expenses, and enhanced efficiency in various sectors. The field of robotics encompasses the conceptualization, creation, and utilization of robots, which can operate autonomously or with minimal human intervention to perform various tasks.

1.4.1 Automation technologies

Industrial automation applies machinery, sensors, and control systems to automate manufacturing and production procedures. The technologies encompassed within this category consist of programmable logic controllers (PLCs), conveyors, and industrial robots. Process automation involves optimizing corporate processes by implementing automated systems that efficiently handle repetitive and rule-based operations. Frequently, it encompasses the utilization of software solutions, such as tools for workflow automation and software for robotic process automation (RPA). Cognitive automation is a technological approach that integrates artificial intelligence, machine learning, and NLP to automate intricate processes that encompass decision-making and data analysis effectively [25]. The technology can effectively process unstructured data and adapt to dynamic settings.

1.4.2 DAM relevance

Automation technology offers several benefits for DAM, including cost reduction, enhanced operational efficiency, scalability, quality control, and a competitive advantage. It replaces manual jobs with machine-based processes, reducing errors and increasing productivity. Automation also maintains consistent standards, enhancing client satisfaction and brand reputation [1]. It also offers improved pricing strategies and superior-quality products or services. Additionally, automation can mitigate safety risks and increase employee safety, making it a valuable tool for DAM in industries with high safety risks.

1.4.3 DAM robotics can streamline processes and reduce costs

Robotics enhances DAM's operational efficiency and cost-effectiveness through many means. These include optimizing efficiency, automating labor-intensive jobs, improving quality control, strengthening inventory management, facilitating flexible production, and promoting workplace safety [3]. Although the initial financial outlay for using robot technology may be substantial, the subsequent advantages of reduced expenses, heightened productivity, and enhanced competitiveness can render it a prudent investment for DAM seeking to flourish in a fiercely competitive market [4].

1.4.3.1 Improved efficiency

Medium-sized enterprises can significantly enhance production efficiency and reduce errors by implementing automation in repetitive operations. Robots demonstrate remarkable precision in assembling, welding, and material handling, providing round-the-clock operational capacity [5]. This mainly benefits DAM with limited operating hours or seasonal demand fluctuations.

1.4.3.2 Reduced labor costs

Labor substitution refers to the process of replacing manual labor with robotic technology, which has the potential to decrease labor costs and mitigate employee turnover. Despite the initial cash infusion, DAM can achieve significant long-term savings by mitigating recruitment and onboarding expenses [6]. Consequently, this can lead to a decreased reliance on hiring more staff.

1.4.3.3 Enhanced quality control

The utilization of robots, which possess distinct attributes such as precision and reproducibility, has a substantial impact on enhancing the quality of products. The notable accuracy exhibited in this context leads to decreased faults and errors, providing DAM with cost-reduction advantages [8]. Advanced sensory technology enables real-time data monitoring, hence facilitating the timely discovery of faults and reducing the necessity for expensive quality assurance procedures. Other DAM services are illustrated in Figure 1.5.

1.4.3.4 Streamlined inventory management

Using robotics in DAM can facilitate the implementation of a simultaneous production strategy, wherein the procurement and delivery of materials and components occur when required. This phenomenon mitigates the necessity for substantial warehouse storage and the accompanying expenses [26]. Automated material handling involves the utilization of robots to manage the movement and manipulation of materials and components efficiently and complete products within a production plant. This practice reduces the likelihood of inventory misuse, damage, or theft.

1.4.3.5 Flexible production

The ability to scale is paramount for DAM to effectively respond to changing market demands and increase its production capabilities. Robotic systems specifically engineered for a wide array of tasks can effectively acquire

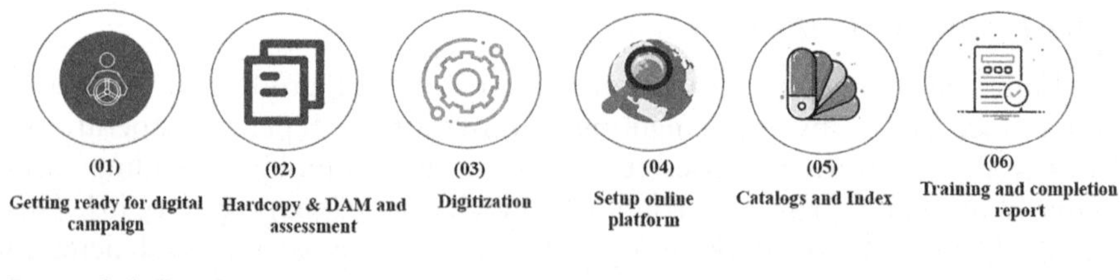

Figure 1.5 DAM services.

supplementary robotic units, augmenting their productivity and reducing the necessity for substantial investments in growth [27].

1.4.3.6 Risk reduction and safety

Workplace Safety: By automating tasks that pose safety risks to human workers, DAM can reduce workplace accidents and the associated costs, for example, workers' compensation claims and regulatory fines [21]. Robotics help DAM manage risks associated with industry-specific compliance and safety standards, thus ensuring compliance can prevent costly legal issues and fines.

1.5 CHALLENGES, LIMITATIONS AND CONSIDERATIONS

This section summarizes common challenges DAM may face when adopting these technologies.

1.5.1 High-speed internet connectivity

Establishing high-speed internet access is decisive in facilitating DAM's widespread adoption of contemporary technologies. Fast and reliable data transfer is a crucial requirement for the effective operation of artificial intelligence models, blockchain networks, and robotics. High-speed internet facilitates uninterrupted connectivity among various devices, enables remote monitoring, and supports utilizing cloud-based services [1]. This, in turn, promotes real-time data exchange and facilitates collaborative efforts.

1.5.2 Cloud computing

Cloud computing allows DAM to access computer resources, storage, and infrastructure in a scalable and cost-effective manner. This technology is valuable in hosting AI models, facilitating blockchain networks, and managing the substantial volumes of data created by robotics [2]. Utilizing remote servers enables DAM to delegate computational activities, diminishing the necessity for substantial on-site infrastructure and decreasing operational expenses.

1.5.3 Edge computing

Edge computing is a paradigm that expands the computational capabilities to the network's periphery, closer to the origin of data generation and the locations where immediate actions are required. The utilization of this technology is of utmost importance in facilitating real-time decision-making within the domains of robotics and AI [3]. DAM can achieve latency reduction, improved reaction times, and enhanced efficiency for their AI and robotics systems by performing data processing on edge devices or servers.

1.5.4 Internet of Things (IoT) sensors

IoT sensors play a crucial role in the acquisition of data from both physical environments and DAM. The sensors detect various characteristics, including temperature, humidity, location, and motion [4]. Integrating IoT data with AI and blockchain systems allows DAM to automate operations, monitor equipment, and obtain valuable insights to facilitate informed decision-making.

1.5.5 Big data analytics

Using big data analytics technologies is imperative to effectively process and extract valuable insights from the extensive data generated by AI, IoT devices, and robotics. DAM has the potential to leverage data analytics in order to recognize patterns, trends, and anomalies [5]. This capability enables them to make data-based decisions, enhance operational efficiency, and enhance customer experiences.

1.5.6 Cybersecurity solutions

Implementing strong cybersecurity measures is of utmost importance to protect DAM's data, AI models, blockchain networks, and robotics systems. These proposed measures safeguard against cyber risks, mitigate the risk of data breaches, and prevent unauthorized access [28]. Cybersecurity measures, including firewalls, encryption, and intrusion detection systems, are implemented to safeguard the security and integrity of essential technological assets.

1.5.7 Interoperability standards

Interoperability standards and protocols play a crucial role in enabling the smooth integration of diverse technologies inside the ecosystems of DAM. These standards facilitate the interchange of data and information among different systems, hence providing efficient communication between AI models, blockchain networks, and robotics components [3]. The concept of interoperability mitigates complexity, enhances efficiency, and streamlines operational procedures.

1.5.8 Digital identity and authentication systems

Integrating AI and robots transforms DAM by providing robust mechanisms that ensure secure access control and data protection. These solutions, supported by digital identity and authentication systems, protect sensitive data and control access to DAM models and devices [8]. Integrating these technologies and robots promotes efficiency, innovation, and competitiveness, promoting DAM growth.

1.6 FUTURE RESEARCH TRENDS

The future outlook for ChatGPT, blockchain, robotics, and AI in DAM is characterized by continued integration and expansion of these technologies into various aspects of business operations. As these technologies advance and become more accessible, DAM will have more opportunities to strategically improve efficiency, competitiveness, and innovation in their respective industries to meet their unique business goals.

1.6.1 DAM chatbots and virtual assistants

The utilization of DAM-driven chatbots has been observed to augment client satisfaction and retention across diverse sectors such as e-commerce, healthcare, and finance. Chatbots can comprehend complex inquiries and contextual nuances, facilitating interactions resembling human-like communication [29]. The company provides customized product suggestions, streamlines transactions, and actively interacts with customers, enhancing customer experience and fostering loyalty.

1.6.2 AI augmentation for DAM

The anticipated trajectory of AI in DAM involves broadening its role to enhance and complement human capabilities. Developments in analytics and predictive capabilities will facilitate the integration of DAM into decision-making processes [20]. This integration will result in improved resource allocation and customized client experiences. This technological advancement is expected to improve operational efficiency, stimulate growth, and uphold competitiveness in rapidly changing market conditions.

1.6.3 Blockchain for supply chain transparency

Blockchain technology can transform supply chain management, specifically for DAM [26]. The decentralized design of the system offers enhanced transparency, traceability, and a reduction in instances of fraudulent operations. Adopting blockchain technology enables DAM to track the authenticity of items effectively, enhance operational effectiveness, and cultivate consumer confidence [8]. The adoption of this approach is motivated by its capacity to reduce expenditures, foster collaboration, and minimize costs.

1.6.4 Robotics in manufacturing and logistics

There is a growing trend among DAM across several industries to incorporate robotics technology into their operations. This adoption has led to increased accessibility and flexibility in utilizing such technology. The adoption encompasses the implementation of precision farming, telemedicine, and warehousing automation [30]. The utilization of robots offers a range of advantages, including enhanced operational efficiency, reduced labor costs, and improved quality control protocols, resulting in heightened production and competitiveness.

1.6.5 DAM-enabled cybersecurity

Medium-sized organizations progressively depend on cybersecurity solutions empowered by AI to detect and alleviate possible risks. These technologies employ sophisticated data processing methodologies to handle substantial amounts of data in real-time effectively, identifying irregularities and potential hazards [5]. Instigating these proactive technologies safeguards DAM data integrity in ongoing commercial activity, thereby bolstering the defense systems of these organizations.

1.6.6 DAM in healthcare and diagnostics

In the healthcare industry, DAM progressively adopts AI to enhance patient outcomes, optimize administrative procedures, and accelerate medical data research. AI algorithms have the potential to assist in the diagnosis of medical disorders, the customization of treatment plans, and the establishment of DAM as leaders in the industry.

1.6.7 DAM-specific AI solutions

There is a growing trend among medium-sized organizations to adopt these trending technologies to meet their specific needs effectively. These solutions, specifically designed to optimize cost-effectiveness and feasibility, are currently being deployed in many domains, such as financial forecasting, human resource administration, marketing automation, and inventory control [3]. Their implementation has resulted in improved operational efficiency and enhanced competitiveness.

1.6.8 Ethical and responsible artificial intelligence

The increasing impact of AI will correspondingly elevate the significance of ethical deliberations. When developing and implementing AI applications, DAM will highly emphasize fairness, transparency, and bias reduction. This encompasses the imperative of assuring non-discriminatory behavior in AI systems, upholding user privacy, and complying with dynamic regulatory frameworks [1]. Implementing AI ethical principles in the field of DAM not only fosters trust among customers and employees but also mitigates the potential for legal and reputational challenges, thereby safeguarding the long-term viability of DAM.

1.7 CONCLUSION

DAMs are progressively utilizing discussed emerging technologies to revolutionize consumer interactions, foster trust, and optimize operational efficiency. AI has the potential to enhance data-driven decision-making processes and bolster cybersecurity measures. DAM operating within the healthcare sector can reap significant advantages from the profound influence of AI. Nevertheless, it is imperative to prioritize ethical concerns to uphold principles of justice and transparency. By embracing these technologies, DAM can strategically position itself for sustained expansion, adaptability, and prominence within an ever-evolving technology landscape. However, it is crucial to prioritize ethical issues to maintain fairness and transparency. By proactively adopting these technologies and ethical practices, DAM can position itself for long-term growth, resilience, and leadership within a constantly shifting technological environment. The future presents opportunities for DAM to adjust to changes and take the lead in driving them, thereby influencing a dynamic and hopeful path forward.

REFERENCES

[1] F. Shao, "New energy industry financial technology based on machine learning to help rural revitalization," *Energy Reports*, vol. 8, pp. 13970–13978, 2022. doi: 10.1016/j.egyr.2022.10.001

[2] W. Shafik, "Toward a more ethical future of artificial intelligence and data science," *In The Ethical Frontier of AI and Data Analysis*, pp. 362–388, 2024. doi. 10.4018/979-8-3693-2964-1.ch022

[3] G. Puthukulam, A. Ravikumar, R. V. K. Sharma and K. M. Meesaala, "Auditors' perception on the impact of artificial intelligence on professional skepticism and judgment in oman," *Universal Journal of Accounting and Finance*, vol. 9, no. 5, 2021. doi: 10.13189/ujaf.2021.090527

[4] W. Shafik, "Introduction to ChatGPT," In *Advanced Applications of Generative AI and Natural Language Processing Models*, vol. 1, pp. 1–25, 2024. doi: 10.4018/979-8-3693-0502-7.ch001

[5] R. Hamadi, H. Ghazzai, H. Besbes and Y. Massoud, "Financial advisor recruitment: A smart crowdsourcing-assisted approach," *IEEE Transactions on Computational Social Systems*, vol. 8, no. 3, pp. 682–688, 2021. doi: 10.1109/TCSS.2020.3040901

[6] A. S. Khan, X. Zhang, S. Lambotharan, G. Zheng, B. Assadhan and L. Hanzo, "Machine learning aided blockchain assisted framework for wireless networks," *IEEE Netw*, vol. 34, no. 5, pp. 262–268, 2020. doi: 10.1109/MNET.011.1900643
[7] W. Shafik, "Navigating emerging challenges in robotics and artificial intelligence in Africa," *Examining the Rapid Advance of Digital Technology in Africa*, vol. 3, pp. 124–144, 2024. doi: 10.4018/978-1-6684-9962-7.ch007
[8] Y. Wang, W. Shafik, J. T. Seong, A. Al Mutairi, M. S. Mustafa and M. R. Mouhamed, "Service delay and optimization of the energy efficiency of a system in fog-enabled smart cities," *Alexandria Engineering Journal*, vol. 84, pp. 112–125, 2023. doi: 10.1016/j.aej.2023.10.034
[9] W. Shafik, "Blockchain-based Internet of Things (B-IoT): Challenges, solutions, opportunities, open research questions, and future trends," *Blockchain-based Internet of Things*, pp. 35–58, 2024. doi: 10.1201/9781003407096-3
[10] M. Polonsky and J. Rotman, "Should artificial intelligent (AI) agents be your co-author? Arguments in favour, informed by ChatGPT," *SSRN Electronic Journal*, 2023. doi: 10.2139/ssrn.4349524
[11] P. S. Mattas, "ChatGPT: A study of AI language processing and its implications," *International Journal of Research Publication and Reviews*, vol. 04, no. 02, 2023. doi: 10.55248/gengpi.2023.4218
[12] L. Caruccio, S. Cirillo, G. Polese, G. Solimando, S. Sundaramurthy and G. Tortora, "Can ChatGPT provide intelligent diagnoses? A comparative study between predictive models and ChatGPT to define a new medical diagnostic bot," *Expert Systems with Applications*, vol. 235, p. 121186. doi: 10.1016/j.eswa.2023.121186
[13] S. AlZu'bi, A. Mughaid, F. Quiam and S. Hendawi, "Exploring the capabilities and limitations of ChatGPT and alternative big language models," In *Artificial Intelligence and Applications*, vol. 2, no. 1, pp. 28–37, 2024. doi: 10.47852/bonviewAIA3202820
[14] S. Biswas, "Role of ChatGPT in computer programming," *Mesopotamian Journal of Computer Science*, 2023. doi: 10.58496/mjcsc/2023/002
[15] A. Htet, S.R. Liana, T. Aung and A. Bhaumik, "ChatGPT in content creation: Techniques, applications, and ethical implications," *In Advanced Applications of Generative AI and Natural Language Processing Models*, pp. 43–68, 2024. doi: 10.4018/979-8-3693-0502-7.ch003
[16] T. Kollmann, L. Kleine-Stegemann, K. de Cruppe and C. Then-Bergh, "Eras of digital entrepreneurship: Connecting the past, present, and future," *Business and Information Systems Engineering*, vol. 64, no. 1, 2022. doi: 10.1007/s12599-021-00728-6
[17] W. Alan Tusa Bagus and D. H. Fudholi, "Klasifikasi Emosi Pada Teks Dengan Menggunakan Metode Deep Learning," *Jurnal Ilmiah Indonesia*, vol. 6, no. 1, 2021.
[18] A. R. Sari, "Skripsi: Perlindungan Hukum Bagi Pemberi Pinjaman Dalam Penyelenggaraan Financial Technology Berbasis Peer To Peer Lending Di Indonesia," *Yogyakarta, Universitas Islam Indonesia*, vol. 6, no. 1, pp. 47–59, 2018.
[19] L. Vinet and A. Zhedanov, "Exploring the importance of financial literacy within the capability approach framework," *J Phys A Math Theor*, vol. 44, no. 8, pp. 220-233, 2011.

[20] B. Lael, "What are we learning about artificial intelligence in financial services?," *J Mater Process Technol*, vol. 1, no. 1, pp. 70–85, 2018
[21] M. U. Tariq, "Revolutionizing health data management with blockchain technology: Enhancing security and efficiency in a digital era," *In Emerging Technologies for Health Literacy and Medical Practice*, pp. 153–175, 2024. doi: 10.4018/979-8-3693-1214-8.ch008
[22] W. Shafik, "A comprehensive cybersecurity framework for present and future global information technology organizations," pp. 56–79, 2023. doi: 10.4018/978-1-6684-9018-1.ch002
[23] L. Zhao, D. Zhu, W. Shafik, S. M. Matinkhah, Z. Ahmad, L. Sharif and A. Craig, "Artificial intelligence analysis in cyber domain: A review," *International Journal of Distributed Sensor Networks*, vol. 18, no. 4. 2022. doi: 10.1177/15501329221084882
[24] W. Shafik and S. M. Matinkhah, "Privacy issues in social web of things," in *2019 5th International Conference on Web Research, ICWR 2019*, Tehran, Iran, pp. 208–214, 2019. doi: 10.1109/ICWR.2019.8765254
[25] V. Merlo, G. Pio, F. Giusto and M. Bilancia, "On the exploitation of the blockchain technology in the healthcare sector: A systematic review," *Expert Systems with Applications*, vol. 213. 2023. doi: 10.1016/j.eswa.2022.118897
[26] W. Shafik, "Making cities smarter: IoT and SDN applications, challenges, and future trends," In *Opportunities and challenges of industrial IoT in 5G and 6G networks*, pp. 73–94, 2023. IGI Global. doi: 10.4018/978-1-7998-9266-3.ch004
[27] A. Sinha, D. Sapra, G. Srivastava, M. Anup and D. Sinwar, "AI-assisted big data analytics for smart healthcare systems," in *Intelligent Internet of Things for Smart Healthcare Systems*, 2023. doi: 10.1201/9781003326182-6
[28] W. Shafik, S. M. Matinkhah and F. Shokoor, "Recommendation system comparative analysis: Internet of Things aided networks," *EAI Endorsed Transactions on Internet of Things*, vol. 8, no. 29, 2022. doi: 10.4108/eetiot.v8i29.1108
[29] H. Abaoğlu, H. Ş. Çiftçi and G. Ekici, "Combined mind-body intervention for premenstrual dysphoric disorder: A randomized controlled trial," *Health Care for Women International*, pp. 1–21. doi: 10.1080/07399332.2024.2314219
[30] W. Shafik, "Artificial intelligence and Blockchain technology enabling cybersecurity in telehealth systems," *Artificial Intelligence and Blockchain Technology in Modern Telehealth Systems*, vol. 1, pp. 285–326, 2023. doi: 10.1049/PBHE061E_ch11

Chapter 2

Transforming healthcare ecosystems with AI-based digital asset management

Ashok, Ravi Kumar Barwal and Neeraj Rohilla

2.1 INTRODUCTION

In the dynamic landscape of healthcare, the integration of Artificial Intelligence (AI) into Digital Asset Management (DAM) is revolutionizing the way we approach patient care, data management, and overall healthcare ecosystems [1]. This groundbreaking synergy holds the potential to transform traditional healthcare models into more efficient, responsive, and patient-centric systems.

At its core, AI-based DAM in healthcare refers to the strategic implementation of advanced algorithms and machine learning models to streamline the management of digital assets, ranging from patient records and diagnostic images to research data and administrative documents. The marriage of AI and DAM not only addresses the ever-growing volume of healthcare data but also enhances the accessibility, security, and utility of these critical assets.

One of the primary advantages of this transformative approach lies in the optimization of clinical workflows. By leveraging AI capabilities, healthcare professionals can automate routine tasks, such as data categorization, analysis, and retrieval. This not only saves valuable time but also minimizes the risk of human errors, ultimately leading to more accurate diagnoses and personalized treatment plans [2].

Furthermore, the integration of AI into DAM facilitates predictive analytics, allowing healthcare providers to anticipate patient needs, track disease trends, and optimize resource allocation. This proactive approach not only enhances preventive care but also contributes to the overall efficiency and cost-effectiveness of healthcare delivery [3].

The security and privacy of patient data are paramount in healthcare, and AI-based DAM systems play a pivotal role in fortifying these aspects. Advanced encryption algorithms and real-time threat detection mechanisms provide a robust defense against cyber threats, ensuring the confidentiality and integrity of sensitive health information [4].

In essence, the amalgamation of AI and DAM in healthcare represents a paradigm shift towards a more intelligent, interconnected, and patient-centric

DOI: 10.1201/9781003590439-2

ecosystem. As we navigate the complexities of modern healthcare, embracing this technological evolution promises not only to enhance operational efficiency but, more importantly, to elevate the standard of care and improve health outcomes for individuals and communities alike.

2.2 LITERATURE REVIEW

Author(s)	*Year*	*Paper title*	*Key findings*
[5]	2003	Public health informatics and information systems	AI-based DAM systems incorporate advanced encryption algorithms. - Real-time threat detection mechanisms enhance cyber security. - Essential for maintaining patient trust and complying with regulatory frameworks such as HIPAA.
[6]	2016	The impact of health information exchange on health outcomes	Emphasizes the ethical implications of AI integration in healthcare DAM. - Calls for transparent and accountable AI algorithms to ensure responsible and equitable healthcare practices. - Stresses the importance of ongoing ethical considerations throughout the development and deployment phases.
[7]	2016	Development and validation of a deep learning algorithm for detection of diabetic retinopathy in retinal fundus photographs	Traditional healthcare data management methods are becoming obsolete. - The surge in data sources necessitates innovative solutions. - AI in DAM automates processes for more effective data handling.
[8]	2011	Data from clinical notes: a perspective on the tension between structure and flexible documentation	AI integration into DAM significantly reduces time spent on routine tasks. - Healthcare professionals can redirect focus toward patient care. - Minimizes the risk of human errors in data management for more accurate diagnoses and treatment plans.
[9]	2005	Improving clinical practice using clinical decision support systems: a systematic review of trials to identify features critical to success	AI-driven DAM enables predictive analytics for healthcare. - Forecasting patient needs, identifying disease trends, and optimizing resource allocation. Proactive approach to patient care through early intervention based on predictive insights.

2.3 Healthcare AI-integrated digital asset management model (HAI-DAM)

Objective: To outline a comprehensive model for seamlessly integrating AI into Digital Asset Management to enhance healthcare ecosystems.

There are multiple features of healthcare AI integrated digital asset management as shown in Figure 2.1.

1. **Data Integration and Centralization**
   - **Objective:** Centralize diverse healthcare data sources for efficient management.
   - **Components:** Electronic Health Records (EHR) [10], Medical Imaging, Genomic Data.
   - **Key Findings:** Integration ensures a unified source for comprehensive patient profiles, facilitating holistic and personalized care.
2. **AI-Driven Automation**
   - **Objective:** Leverage AI algorithms to automate routine tasks and data processes.
   - **Components:** Natural Language Processing (NLP), Machine Learning (ML) [11].
   - **Key Findings:** Significant reduction in time spent on data categorization, analysis, and retrieval, leading to improved workflow efficiency.
3. **Predictive Analytics for Proactive Healthcare**
   - **Objective:** Utilize AI for predictive analytics to forecast patient needs and identify disease trends.

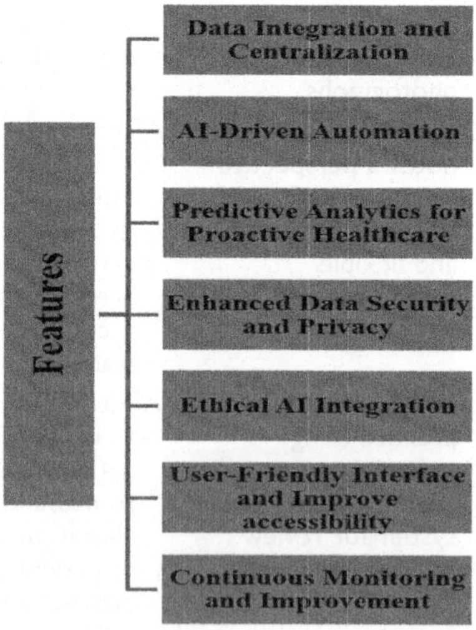

Figure 2.1 Features of healthcare AI integrated digital asset management.

- **Components:** Machine Learning Models, Predictive Modeling.
- **Key Findings:** Proactive healthcare strategies, early intervention, and optimized resource allocation for improved patient outcomes.

4. **Enhanced Data Security and Privacy**
   - **Objective:** Implement advanced cyber security measures within AI-based DAM systems.
   - **Components:** Encryption Algorithms, Real-time Threat Detection [12].
   - **Key Findings:** Ensures robust defense against cyber threats, maintaining patient trust and compliance with regulatory frameworks.
5. **Ethical AI Integration**
   - **Objective:** Embed ethical considerations in the development and deployment of AI algorithms.
   - **Components:** Ethical AI Guidelines, Transparency Measures.
   - **Key Findings:** Promotes responsible and equitable healthcare practices, addressing potential biases and ensuring patient data integrity.
6. **User-Friendly Interface and Accessibility**
   - **Objective:** Design a user-friendly interface for seamless accessibility.
   - **Components:** User Experience (UX) Design, Accessibility Features.
   - **Key Findings:** Improves healthcare professionals' engagement with the system, fostering better utilization and adoption.
7. **Continuous Monitoring and Improvement**
   - **Objective:** Establish mechanisms for continuous monitoring and refinement.
   - **Components:** Monitoring Tools, Feedback Loops.
   - **Key Findings:** Enables ongoing optimization, addressing emerging challenges and incorporating user feedback for continuous improvement.

This model, HAI-DAM, envisions a cohesive and intelligent integration of AI into Digital Asset Management in healthcare. By addressing key aspects such as data integration, automation, predictive analytics, security, ethics, user accessibility, and continuous improvement, this model aims to redefine healthcare ecosystems for enhanced patient care and overall efficiency.

2.4 ADVANTAGES OF HEALTHCARE AI-INTEGRATED DIGITAL ASSET MANAGEMENT MODEL (HAI-DAM)

1. **Efficient Data Management**
   HAI-DAM streamlines the management of diverse healthcare data sources, ensuring centralized and efficient handling of electronic health records, medical imaging, genomic data, and other critical assets.

2. **Workflow Optimization**
   The model employs AI-driven automation to significantly reduce the time healthcare professionals spend on routine tasks, leading to optimized clinical workflows and allowing more focus on patient care.
3. **Predictive Analytics for Proactive Care**
   Leveraging predictive analytics, HAI-DAM enables healthcare providers to anticipate patient needs, identify emerging disease trends, and adopt a proactive approach to healthcare, ultimately improving patient outcomes.
4. **Enhanced Data Security**
   With advanced encryption algorithms and real-time threat detection, HAI-DAM ensures robust cyber security, safeguarding patient data and maintaining trust by complying with regulatory frameworks such as HIPAA.
5. **Ethical AI Integration**
   The model emphasizes ethical considerations, promoting transparency and accountability in AI algorithms. This ethical integration helps mitigate biases and ensures fair and equitable healthcare practices.
6. **Personalized Patient Care**
   By centralizing and analyzing comprehensive patient profiles, HAI-DAM facilitates personalized care strategies. Healthcare professionals can make more informed decisions tailored to individual patient needs.
7. **Improved Resource Allocation**
   Predictive analytics enables optimized resource allocation, helping healthcare organizations allocate personnel, equipment, and other resources more effectively based on anticipated patient needs and trends.
8. **User-Friendly Interface**
   HAI-DAM is designed with a user-friendly interface, enhancing accessibility for healthcare professionals. This intuitive design fosters better engagement and utilization of the system.
9. **Cost Efficiency**
   Through workflow optimization and improved resource allocation, HAI-DAM contributes to overall cost efficiency in healthcare delivery. The model helps organizations make informed decisions that maximize the value of their resources.
10. **Continuous Improvement**
    The model incorporates mechanisms for continuous monitoring and refinement, allowing for ongoing optimization. This adaptability ensures that HAI-DAM remains responsive to emerging challenges and user feedback, contributing to sustained improvements in healthcare ecosystems.

2.5 LIMITATIONS OF HEALTHCARE AI-INTEGRATED DIGITAL ASSET MANAGEMENT MODEL (HAI-DAM)

1. **Data Privacy Concerns**
   Despite advanced encryption, there may still be concerns about the privacy of sensitive health data, especially considering the increasing frequency of cyber threats and data breaches.
2. **Implementation Costs**
   The initial implementation of the HAI-DAM model, including the integration of AI technologies and cyber security measures, may pose significant financial challenges for healthcare organizations.
3. **Technological Dependence**
   The model's effectiveness relies on the continuous functionality of AI technologies. Any technical glitches, system failures, or dependencies on external technologies could disrupt healthcare operations.
4. **Potential Bias in AI Algorithms**
   Despite ethical considerations, AI algorithms may inadvertently inherit biases present in training data, potentially leading to biased decision-making and healthcare disparities.
5. **User Resistance and Training**
   Healthcare professionals may resist adopting new technologies, and training programs may be necessary to ensure efficient utilization, adding an additional burden during the implementation phase.
6. **Limited Generalization**
   The effectiveness of AI algorithms in predictive analytics may be limited in certain healthcare contexts and the model's generalization to diverse patient populations or rare medical conditions may be challenging.
7. **Regulatory Compliance Challenges**
   Evolving healthcare regulations and compliance standards may pose challenges in keeping HAI-DAM consistently aligned with changing legal requirements, potentially leading to non-compliance issues.
8. **Data Integration Complexity**
   Integrating various data sources into a cohesive system may be complex, especially when dealing with legacy systems or heterogeneous data formats across different healthcare institutions.
9. **Patient Acceptance and Trust**
   Patients may be apprehensive about the use of AI in healthcare, and building and maintaining trust in the security and ethical handling of their data is crucial for successful implementation.
10. **Overemphasis on Technology**
    There is a risk of overemphasizing technological solutions without addressing broader systemic issues within healthcare, potentially overlooking the human and social aspects of patient care.

2.6 FUTURE SCOPE

1. **Efficient Data Management**: HAI-DAM will revolutionize healthcare data management by providing a centralized, organized, and easily accessible repository for digital assets. This includes medical images, patient records, research data, and more.
2. **Enhanced Diagnostics**: The integration of AI in DAM allows for advanced image recognition and analysis. This means quicker and more accurate diagnostics, aiding healthcare professionals in making informed decisions.
3. **Personalized Treatment Plans**: AI algorithms within HAI-DAM can analyze vast datasets to identify patterns and trends related to specific health conditions. This information can be used to tailor personalized treatment plans for patients, considering individual variations in response to treatments.
4. **Predictive Analytics**: By leveraging machine learning, HAI-DAM can predict potential health issues based on historical data. This proactive approach enables early intervention and preventive measures, ultimately improving patient outcomes.
5. **Research Advancements**: The centralized nature of HAI-DAM facilitates collaborative research efforts. Researchers can access a wealth of diverse healthcare data, accelerating the pace of medical discoveries and innovations.
6. **Interoperability and Integration**: HAI-DAM can bridge the gap between different healthcare systems and devices. This interoperability ensures seamless sharing of data across various platforms, promoting a holistic approach to patient care.
7. **Cost-Efficiency**: The streamlined processes and automation brought about by HAI-DAM can lead to cost savings in healthcare operations. This includes reduced administrative overhead, quicker decision-making, and optimized resource utilization.
8. **Improved Patient Engagement**: With secure access to their health data, patients become active participants in their care. HAI-DAM can support patient engagement tools, such as health apps and portals, fostering a collaborative relationship between healthcare providers and patients.
9. **Regulatory Compliance**: HAI-DAM can be designed to adhere to healthcare regulations and privacy standards. This ensures the secure handling of sensitive patient information, maintaining trust and compliance with legal requirements.
10. **Continuous Learning and Adaptation**: The AI components of HAI-DAM can continuously learn from new data inputs, evolving to stay current with the latest medical advancements. This adaptability is crucial in a dynamic healthcare landscape.

2.7 CONCLUSION

In conclusion, the Healthcare AI-Integrated Digital Asset Management Model (HAI-DAM) stands at the forefront of a healthcare revolution. Its potential to reshape the landscape of medical data management, diagnostics, and patient care is nothing short of groundbreaking. The convergence of artificial intelligence with digital asset management not only enhances the efficiency of healthcare operations but also paves the way for unprecedented advancements in personalized medicine, research, and predictive analytics.

As HAI-DAM continues to evolve, it holds the promise of not just managing data but extracting meaningful insights from it. The seamless integration of AI algorithms empowers healthcare professionals with tools to make more accurate and timely decisions, ultimately leading to improved patient outcomes.

Moreover, the collaborative and interoperable nature of HAI-DAM fosters a synergistic environment where healthcare providers, researchers, and patients can actively engage in the healthcare journey. The model's adherence to regulatory standards ensures the secure handling of sensitive information, addressing the ethical concerns associated with healthcare data.

In essence, HAI-DAM is not merely a technological innovation; it represents a paradigm shift in how we approach healthcare. Its potential to drive continuous learning, adaptability, and cost-efficiency positions it as a cornerstone in the future of healthcare delivery. As we embrace the era of smart healthcare, HAI-DAM emerges as a beacon of progress, promising a future where data becomes a powerful ally in the pursuit of better health and well-being for all.

REFERENCES

1. Pan, J., & Plaisant, C. (2009). "Beyond Search: Image Classification and Querying in the Medical Image Resource Center." *Journal of the American Society for Information Science and Technology*, 60(3), 555–566.
2. Pankowska, M., Seinstra, F. J., & Dijkstra, J. (2018). "Exploring Digital Asset Management in Healthcare Research: A Case Study." In *European Conference on Information Systems (ECIS)*.
3. Hosny, A., Parmar, C., Quackenbush, J., & Schwartz, L. H. (2018). "Artificial Intelligence in Radiology." *Nature Reviews Cancer*, 18(8), 500–510.
4. Rana, S. K., Rana, A. K., Naz, S. M. I., Singh, A. K., & Bhushan, B. (2023, May). "Supply Chain Strategies Supported by Blockchain Technology: Open Problems and Challenges." In *2023 International Conference on Advancement in Computation & Computer Technologies (InCACCT)* (pp. 1–6). IEEE.
5. O'Carroll, P. W., Yasnoff, W. A., Ward, M. E., & Ripp, L. H. (2003). *Public health informatics and information systems*. Springer.
6. Kim, Y., & Yoon, Y. (2016). "The Impact of Health Information Exchange on Health Outcomes." *Applied Clinical Informatics*, 7(4), 1050–1061.

7. Gulshan, V., Peng, L., Coram, M., Stumpe, M. C., Wu, D., Narayanaswamy, A., ... & Webster, D. R. (2016). "Development and Validation of a Deep Learning Algorithm for Detection of Diabetic Retinopathy in Retinal Fundus Photographs." *JAMA*, 316(22), 2402–2410.
8. Rosenbloom, S. T., Denny, J. C., Xu, H., Lorenzi, N., & Stead, W. W. (2011). "Data From Clinical Notes: A Perspective on the Tension Between Structure and Flexible Documentation." *Journal of the American Medical Informatics Association*, 18(2), 181–186.
9. Kawamoto, K., Houlihan, C. A., Balas, E. A., & Lobach, D. F. (2005). "Improving Clinical Practice Using Clinical Decision Support Systems: A Systematic Review of Trials to Identify Features Critical to Success." *BMJ*, 330(7494), 765.
10. Rana, S. K., Rana, S. K., Rana, A. K., Nisar, K., Soomro, T. R., & Nisar, S. (2022, November). "A Survey on Blockchain Technology Supported Approaches for Healthcare System, Open Issues and Challenges." In *2022 14th International Conference on Mathematics, Actuarial Science, Computer Science and Statistics (MACS)* (pp. 1–7). IEEE.
11. Dinesen, B., Nonnecke, B., Lindeman, D., Toft, E., Kidholm, K., Jethwani, K., ... & Southard, J. A. (2016). "Personalized Telehealth in the Future: A Global Research Agenda." *Journal of Medical Internet Research*, 18(3), e53.
12. Schulz, S., Balkanyi, L., Cornet, R., Bodenreider, O., & van Mulligen, E. (2011). "From Concept Representations to Ontologies: A Paradigm Shift in Health Informatics?" *Healthcare Informatics Research*, 17(4), 235–242.

Chapter 3

Image encryption decryption using hybrid cryptography

Kuldeep Vayadande

3.1 INTRODUCTION

With the increasing need for data security, image encryption has become a crucial aspect of modern information technology. Image encryption, a method of transforming an image into a secret code to protect its confidentiality, has become an essential tool to safeguard valuable data. However, traditional image encryption techniques have been rendered less effective by the rapid advancements in technology, leaving data vulnerable to cyber-attacks. Therefore, there is an urgent need to develop a new and more robust image encryption system to address these issues and enhance the security of sensitive information.

Image encryption and decryption are crucial in protecting sensitive data from unauthorized access, tampering, or theft. Unfortunately, new technological developments have rendered conventional encryption methods outmoded and attack-prone. The time needed to defeat a ciphertext has considerably decreased due to advancements in technology. Additionally, the various defects in separate systems frequently affect their effectiveness and performance. Thus, it is more crucial than ever to have a system that takes into account the trade-off between privacy and speed when utilizing cryptography. The transmission of digital images over the internet has become a common practice in today's digital world. This transmission, however, exposes the images to various security threats such as cyber-attacks, hacking, and malware. An image encryption system provides a solution to these challenges, ensuring the security and confidentiality of transmitted images. The ability to secure digital images through encryption and decryption has become more important than ever before in this digital age, characterized by the proliferation of cyber threats.

The study suggests combining the symmetric cryptography techniques AES and Blowfish with the asymmetric cryptography system RSA to solve the exchange between security and performance. Often employed for encrypting data on the TLS Layer, RSA requires two unique keys to encode and decode data. The quickest technique is Blowfish, but the most effective and reliable encryption is offered by AES. To guarantee key safety, the

DOI: 10.1201/9781003590439-3

suggested algorithm encrypts data three times using the three distinct methods. Combining several algorithms can effectively resolve the privacy and performance trade-off by addressing the shortcomings of each one's individual applications. The suggested system may be used to offer safe data encryption and decryption capabilities in a wide range of applications and operations.

The introduction lays out the context and significance of the research. Following this, the related work section reviews existing literature and methodologies pertinent to the study. A comparison table is provided to highlight key differences among cryptographic algorithms used in similar contexts. The section on cryptographic algorithms delves into the specifics of the chosen algorithm and its rationale for the proposed system. Methodology and implementation detail the approach taken to conduct the research and develop the proposed system. Subsequently, the proposed system is introduced, elucidating its design and functionality. The performance of the suggested system is graphically contrasted with that of current solutions in a comparison graph, which leads to the results and discussions section where the conclusions are examined. The scope of research outlines potential applications and extensions of the proposed system, followed by a conclusion summarizing key insights. Lastly, future scope suggests avenues for further research, and references list the works cited throughout the paper.

3.2 RELATED WORKS

In the realm of cloud security, various cryptographic techniques have been explored to enhance data protection. K. R. Sajay et al. [1] proposed a hybrid encryption algorithm that integrates both symmetric and asymmetric encryption methods, specifically designed to secure cloud data. This approach effectively balances security and performance, ensuring that cloud-stored data remains protected against unauthorized access.

An extensive examination of hybrid data security techniques designed for cloud systems was carried out by V. K. Soman and V. Natarajan [2]. Their work compares the efficiency and effectiveness of various cryptographic schemes, concluding that hybrid algorithms, which combine multiple encryption techniques, offer a superior level of security while maintaining computational efficiency.

B. Hammi et al. [3] presented a lightweight ECC-based authentication method designed specifically for the Internet of Things (IoT). This technique makes advantage of elliptic curve cryptography (ECC) to provide strong security protections while addressing the particular limitations of Internet of Things devices, such as limited processing power and storage.

In a similar vein, P. Kumar Mahapatra et al. [4] developed a security model aimed at preserving the privacy of images stored in the cloud. Their approach utilizes a combination of cryptographic techniques to ensure that sensitive image data is protected during both storage and transmission, addressing the growing need for privacy in cloud computing environments.

R. Hentschel et al. [5] examined the current challenges faced by cloud service providers in Germany, highlighting the importance of adopting advanced security measures to protect cloud infrastructure. Their study underscores the need for robust encryption techniques to mitigate risks associated with data breaches and other cyber threats in cloud computing.

Finally, F. K. Mupila and H. Gupta [6] proposed an innovative authentication model designed to enhance cloud security. Their model integrates multiple layers of encryption and authentication mechanisms to safeguard data in cloud environments, providing a comprehensive solution to the challenges of cloud security.

These studies collectively highlight the evolving landscape of cloud security, emphasizing the importance of developing and implementing advanced cryptographic techniques to protect sensitive data in an increasingly interconnected digital world.

Table 3.1 shows various techniques, their advantages and disadvantages.

Table 3.1 Comparison table [1–11]

Sr. No.	*Technique*	*Advantages*	*Disadvantages*
1	An unorganized key-based scheme for encryption and decryption of images	Low hardware cost, excellent security, low computational complexity, and great hardware use efficiency.	Hardware placement and routing issues.
2	Lossless Image Compression and Encryption Using SCAN	confidentiality and produces a large variety of space filling curves or scanning pathways. efficiency of hardware consumption.	Less data holding capacity.
3	Hybrid model using AES and DES	Increases the nonlinear characteristics of the basic AES and increases dissemination when it is coupled with DES.	Increased time complexity as compared to the algorithms used individually.
4	Hybrid model using RSA and AES	Increased security and Integrity.	Slow processing, hence increased time complexity.
5	Symmetric key block cipher using AES	A rapid rate of transmission with a restricted frequency, redundancies, bulk capacities and pixel correlation.	For some applications that demand a greater level of security, AES's fixed key size of 128 bits might not be sufficient.

(*Continued*)

Table 3.1 (Continued) Comparison table [1–11]

Sr. No.	*Technique*	*Advantages*	*Disadvantages*
6	A Comparative Analysis of Public Key Cryptography using RSA & ECC	Together these are more efficient than RSA and resistant to quantum computing attacks.	It is susceptible to timing assaults and differential power analysis, two kinds of side-channel attacks.
7	Combined Relative Displacement (RDC) and Dynamic Base Transformation (DBTC) Symmetric Key Cryptography (CRDDBT)	Potentially increases the whole encryption scheme's security.	The system may be vulnerable to attacks that target the specific encryption methods used
8	Using symmetric DES for encryption and decryption	DES was widely adopted and implemented in various systems and applications, making it easy to integrate into existing systems along with its speed.	DES is known to be vulnerable to differential cryptanalysis, which is a method of cryptanalysis that can be used to recover the key used in the encryption.
9	Algorithm based on chaotic economic model	Chaotic systems are known to be highly sensitive to initial conditions and small changes in the system, making it difficult for an attacker to predict the behavior of the system and break the encryption.	There isn't a widely accepted standard for chaotic economic model-based encryption, which can make it difficult to compare and evaluate different algorithms.
10	DES, AES and Blowfish for Image Encryption & Decryption	A hybrid encryption system combines the strengths of multiple algorithms, increasing the difficulty with which attackers must crack the encryption.	Increased vulnerabilities that can potentially be used to break the encryption, making the algorithm less secure than some other options.
11	Encryption and Decryption using block-based blowfish algorithm.	It uses a variable-length key, which can be any length between 32 and 448 bits. This allows for a high level of security, as longer keys are more difficult to break.	Some software and hardware devices may not include built-in support for Blowfish, making it more difficult to use in certain situations.

3.3 CRYPTOGRAPHIC ALGORITHMS USED

3.3.1 Advanced Encryption Standard (AES)

Developed to offer dependable data encryption and decryption, the Advanced Encryption Standard (AES) is a type of symmetric block cipher technology. Among the many different applications that heavily use it are digital signatures, online banking, and wireless communication [12, 13]. To begin with, the plaintext is split into fixed-size blocks (usually 128 bits long) so that the AES algorithm can work. Substitution, permutation, and mixing are among the several operations that are subsequently carried out on every block. The secret key used for these operations is shared by the encrypted communication's sender and recipient. The key length, as shown in Figure 3.1, can have one of three values: 128 bits, 192 bits, or 256 bits. This determines the encryption security level. AES has been utilized since. Standardized by the National Institute of Standards and Technology (NIST) in the United States, the AES algorithm withstands a range of cryptographic assaults, such as linear and differential cryptanalysis. Sensitive data must be protected in modern computer settings, and the AES algorithm is one of the most secure encryption techniques on the market. The AES Block diagram is shown in Figure 3.1.

3.3.2 Blowfish algorithm

These operations consist of a key-dependent mix-ing function, substitution, and permutation. Blowfish employs a secret key that is shared by the sender and recipient of the encrypted message, as shown in Figure 3.2, and can have a length of 32 to 448 bits. Blowfish's changeable key length is one of its distinctive features; it allows it to be extremely versatile and adaptive to various security requirements [14, 15]. Its speed is another advantage that makes it a useful encryption technique for a wide range of real-world uses. Blowfish is a trustworthy and safe encryption technique as it has resisted several attacks and has been the subject of extensive cryptographic examination. All things considered, the Blowfish algorithm is a great option for anyone looking for a quick and safe encryption technique. As seen in Figure 3.2, Blowfish Block Diagram

3.3.3 RSA algorithm

The RSA algorithm is a public-key cryptography technique developed by Ron Rivest, Adi Shamir, and Leonard Adleman. It is widely used in modern computer security to provide a secure connection and protect personal data [16, 17]. Two keys are needed for the RSA technique: a public key and a private key, as shown in Figure 3.3. The public key is used to encrypt the data, while the private key is used to decrypt it. The system's security is

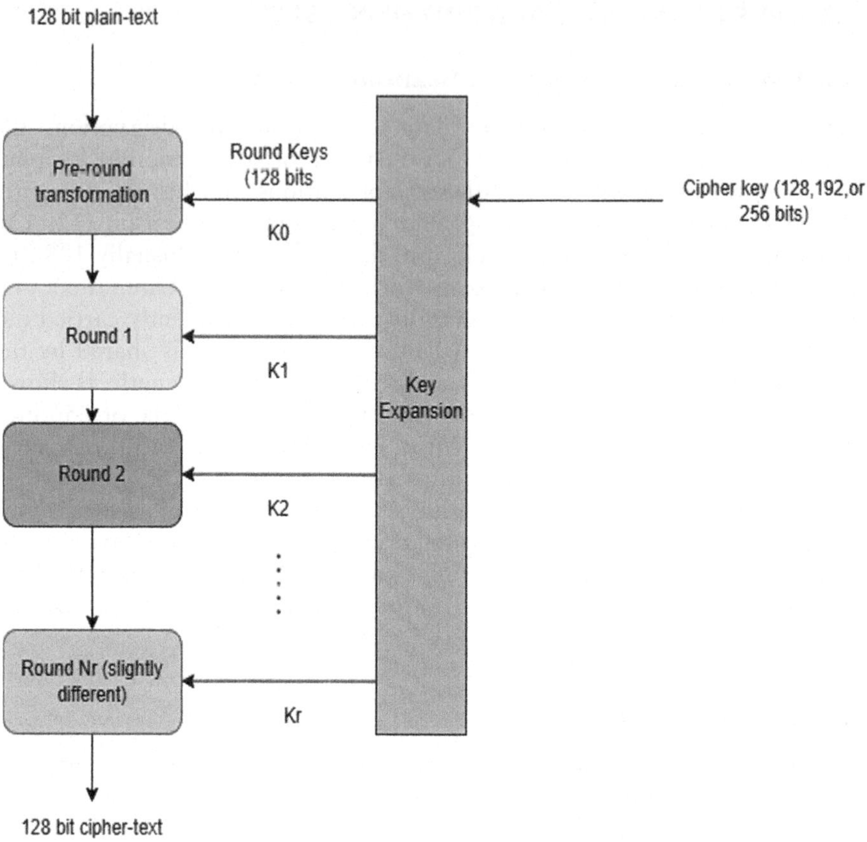

Figure 3.1 AES block diagram.

based on the notion that large integers cannot be computationally factored into their prime factors. The RSA algorithm consists of four main stages: key generation, encryption, decryption, and distribution. Taking two huge prime integers, p and q, then multiplying them together to get n=p*q is the method of producing keys.Starting with n, we obtain the value e, which is relatively prime to (p-1)(q-1). The sources of the private key are P, Q, and the public key. During the encryption procedure, the sender encrypts the message using the recipient's public key. During the decryption process, the recipient uses their private key to decrypt the communication. The RSA method is widely used in applications such as digital signatures, secure email exchange, and SSL/TLS. Considered to be among the most secure encryption methods available, it is frequently employed as an essential instrument in modern computer settings to protect sensitive information. The p, q, and public key contexts to protect sensitive data. p, q, and the public key.

Figure 3.3 shows RSA Block Diagram.

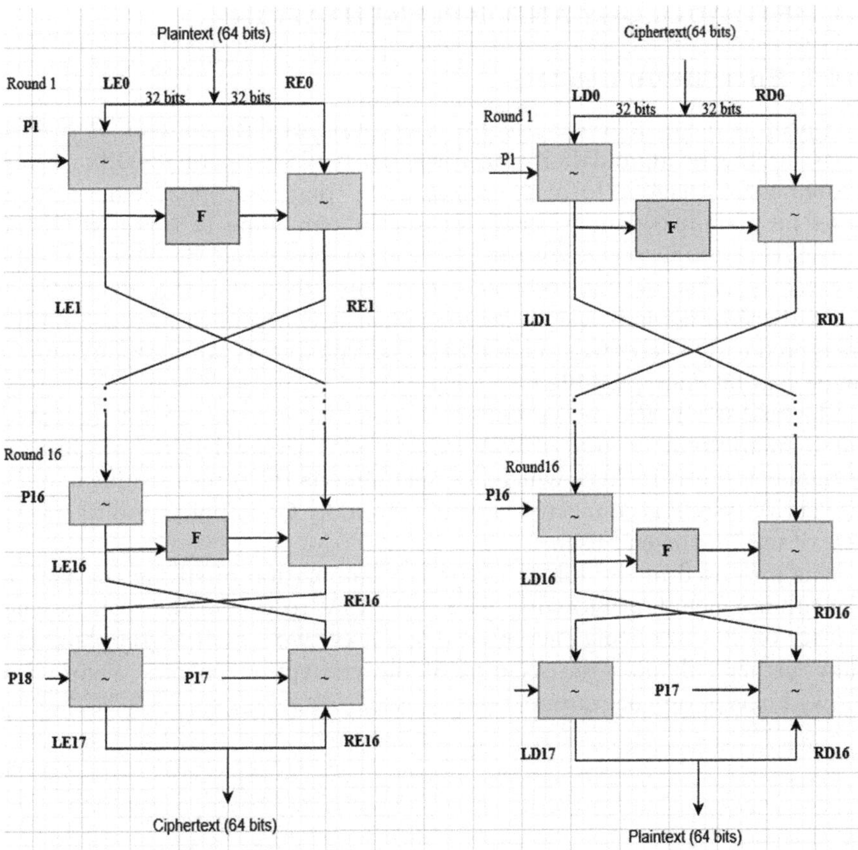

Figure 3.2 Blowfish block diagram.

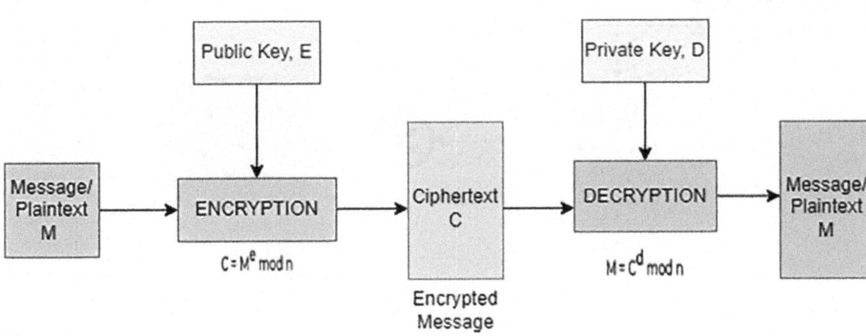

Figure 3.3 RSA block diagram.

3.4 METHODOLOGY AND IMPLEMENTATION

3.4.1 Encryption process

The proposed methodology for the encryption of data consists of several steps. Firstly, the input data is transformed to a 128-bit plain text after being treated as an image in formats such as .jpeg, .png, and .jpg. Then, two sets of 64-bit plain text data are created from the remaining 128-bit text [18].

The next step is to encrypt the 64-bit plain data using the Blowfish algorithm to produce 64-bit encrypted text. After that, the AES algorithm is given the 64-bit plain text as input, which it encrypts to produce 64-bit encrypted text. These two sets of 64-bit encrypted texts are combined into a single 128-bit encrypted file.

The final step in the encryption process is to subject the 128-bit encrypted file to an additional round of RSA encryption. This involves the production of public and private keys which are then sent via the transmission channel. The entire system is connected via a user-friendly GUI application to ensure ease of use for the end-user.

This proposed methodology utilizes a combination of strong encryption algorithms, including Blowfish, AES, and RSA, to ensure the security and privacy of the data being encrypted. The use of two sets of 64-bit plain text data further enhances the security of the encryption process. Figure 3.4 shows Encryption Diagram.

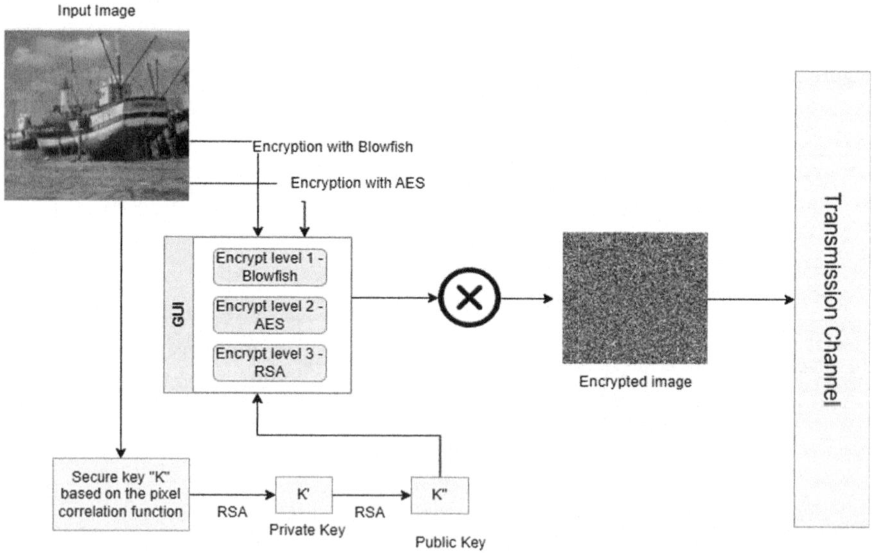

Figure 3.4 Encryption diagram.

3.4.2 Decryption process

The proposed methodology for decryption of data is based on the RSA algorithm, which decrypts the data into sets of 128 bits using both public and private keys. The encrypted data, which is 128-bit long, is split into two 64-bit data sets. These two data sets are then subsequently applied to the Blowfish method to obtain two decrypted sets of 64 bits [19, 20].

After that, the following data sets are further processed using the AES technique to produce two sets of decrypted 64-bit data. These two 64-bit decrypted data sets are then combined to form a single 128-bit data set. The decrypted data can be saved in the original or any other format compatible with the system.

The decryption process is connected via a user-friendly GUI application to make it easy to use for the users. The RSA algorithm provides high security for the decryption process, as it uses both public and private keys to decrypt the data [21, 22]. The Blowfish method and AES technique add an extra layer of security to ensure that the decrypted data is safe from unauthorized access.

The proposed decryption methodology can be implemented in various applications that require secure communication and data transfer, including online transactions and cloud computing. The user-friendly GUI application makes it easy for the users to decrypt the data and save it in the desired format. Figure 3.5 shows Decryption Flowchart

3.4.3 Proposed system

3.4.3.1 *Cascading manner of keys encryption for the data*

The proposed encryption system is composed of three encryption layers, a key generator, and a list of keys. The system encrypts the plaintext using the

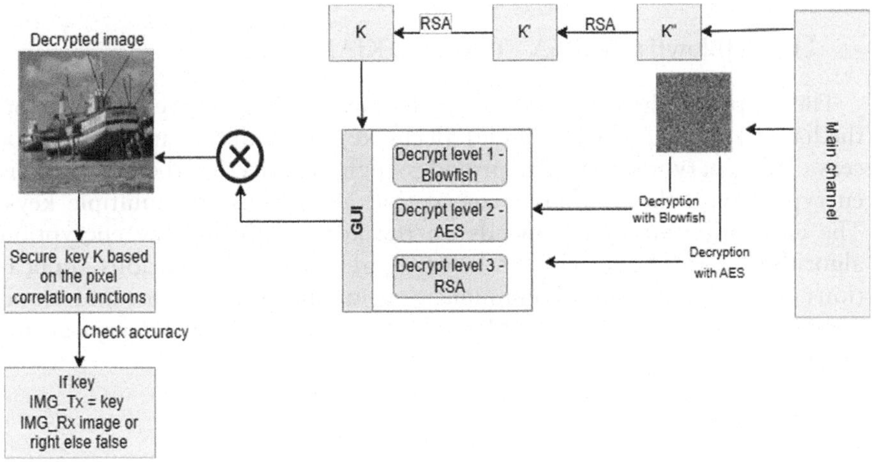

Figure 3.5 Decryption flowchart.

Blowfish algorithm with a 32-bit, 64-bit, or 128-bit key, KIBlowfish. The key is generated by the key generator and is used for Blowfish encryption. The resulting ciphertext, CI1, is obtained by applying the Blowfish encryption to the plaintext. The generated key is then appended to the list of keys, LI. The list of keys LI is updated using the XOR operation to include the new key.

$$\text{CI1} = \text{Blowfish}\left(\text{Plaintext} = \text{PI}, \text{Key} = \text{KIBlowfish}\right)$$

$$\text{L I} = [\,] \oplus \text{KIBlowfish}$$

After the first encryption layer, the ciphertext, CI1, is encrypted using RSA encryption with a 1024-bit or 2048-bit public key, KRSA-Public, generated by the key generator. The private key, $\text{KI}_{\text{RSA-Private}}$, is also generated for decryption. The private key is appended to the list of keys, LI. The resulting ciphertext, C2I, is obtained by applying RSA encryption to the ciphertext, CI1. The list of keys, LI, is updated using the XOR operation to include the new private key.

$$\text{C2I} = \text{RSA}\left(\text{Plaintext} = \text{CI1}, \text{Key} = \text{KIRSA\_Public}\right)$$

$$\text{LI} = \left[\text{KIBlowfish}\right] \oplus \text{KIRSA} - \text{Private}$$

Finally, the ciphertext, C2, is encrypted using AES-128 encryption with the 128-bit key, KAES, generated by the key generator. The key is appended to the list of keys, LI. The final encrypted ciphertext, CI, is obtained by applying AES encryption to the ciphertext, C2I. The list of keys, LI, is updated using the XOR operation to include the new AES key.

$$\text{Ciphertext, CI} = \text{AES}\left(\text{Plaintext} = \text{C2I}, \text{Key} = \text{KIAES}\right)$$

$$\text{LI} = \left[\text{KIBlowfish, KIRSA} - \text{Private}\right] \oplus \text{KIAES}$$

The output of the proposed encryption system is the ciphertext, CI, and the list of keys, LI, which includes all the keys used in the encryption process. This encryption system provides high security, as the plaintext is encrypted with three different encryption algorithms and multiple keys. The encryption strength depends on the key length and the encryption algorithm used. The system can be easily implemented in various applications that require secure communication, online transactions, and cloud computing. Figure 3.6 shows Cascading Manner of Keys Encryption for Data

$$\text{Ciphertext, CI}$$

$$\text{List of keys, LI} = \left[\text{KIBlowfish, KIRSA} - \text{Private, KIAES}\right]$$

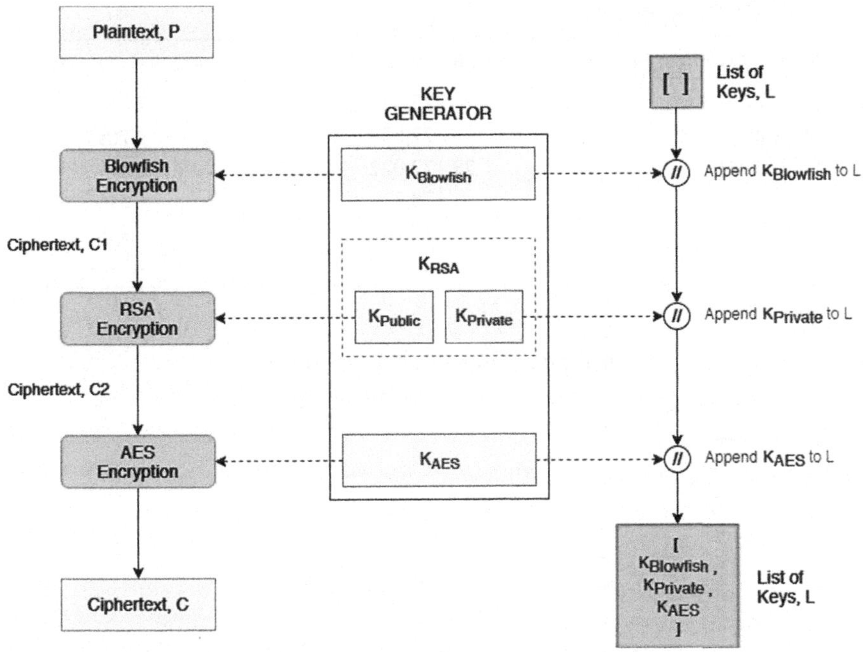

Figure 3.6 Cascading manner of keys encryption for data.

3.4.4 Graph of comparison

The comparison of various images are tested on the proposed system and the time taken for each is recorded which is being showcased below.

3.4.4.1 Comparison for encryption time

Table 3.1 presents data on the length of cipher text, plain text, and encryption time for various instances of encryption. The encryption time for each instance ranges from 1.3 to 69 seconds, with the cipher text and plain text lengths varying accordingly. The data represented in Figure 3.5 suggests that longer plain text and cipher text lengths result in longer encryption times. Overall, the table provides insight into the performance of encryption algorithms and the tradeoffs between security and processing speed. Table 3.2 shows Encryption time log and Figure 3.7 shows Comparison on Encryption Time and Cipher Text length

3.4.4.2 Comparison of decryption time

The table in Table 3.2 provides information on the decryption time required for various cipher and plain text lengths. The decryption time is measured in seconds and ranges from 1.06s to 40.0s. The cipher text lengths range from

Table 3.2 Encryption time log

Cipher text length (bit)	*Plain text length (bit)*	*Encryption time (sec)*
304,672	225,930	2.26s
30,982	22,638	2.70s
39,312,416	29,177,022	69s
235,040	174,330	1.3s
37,920	27,750	1.34s
304,672	225,930	1.9s

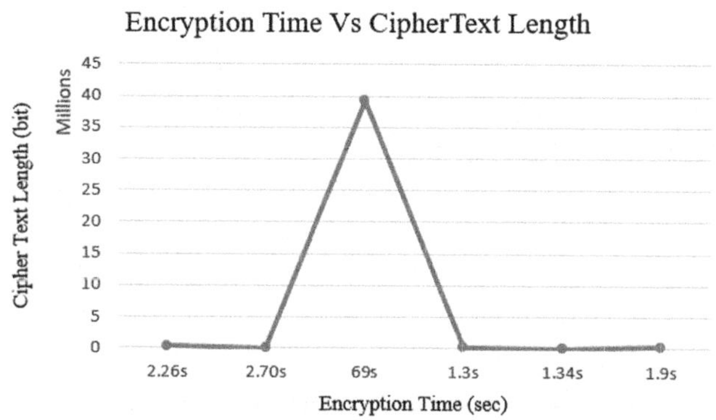

Figure 3.7 Comparison on encryption time and cipher text length.

30,982 bits to 39,312,416 bits, while the corresponding plain text lengths range from 22,638 bits to 29,177,022 bits. The decryption time generally increases as the cipher and plain text lengths increase like seen in the Figure 3.5. However, there are exceptions as seen in the case of the 304,672-bit cipher text and 225,930-bit plain text, which had different decryption times of 1.26s and 1.87s respectively. Overall, the table provides useful information for evaluating the efficiency of different decryption methods. Table 3.3 shows Decryption time log. Figure 3.8 shows graph for Decryption Time Vs Ciphertext Length

Table 3.3 Decryption time log

Cipher text length (bit)	*Plain text length (bit)*	*Decryption time (sec)*
304,672	225,930	4.26s
30,982	22,638	10.70s
39,312,416	29,177,022	88.0s
235,040	174,330	10.97s
37,920	27,750	8.06s
304,672	225,930	5.87s

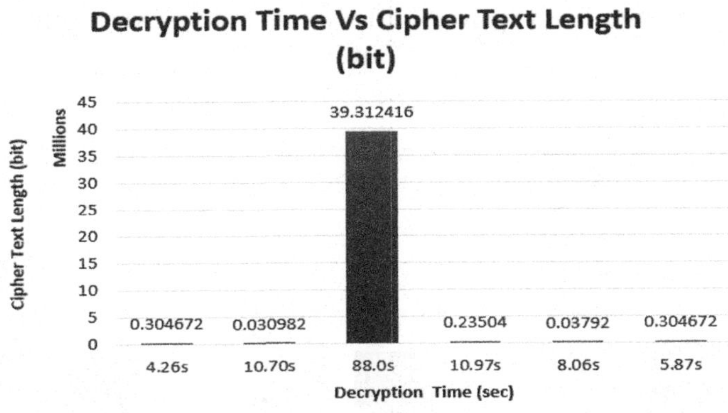

Figure 3.8 Decryption time Vs ciphertext length.

3.4.5 Performance comparison with other existing models

The table in Table 3.3 presents the average encryption and decryption time in seconds for three different image encryption systems: Image-Hybrid Crypto System (Blowfish-RSA-AES), Standalone Image Blowfish System, and Combination of RSA and AES System. The Image-Hybrid Crypto System takes the longest time for decryption, with an average of 500.8 seconds which makes it the hardest among all others to break, while the Standalone Image Blowfish System is the fastest, taking only 125.6 seconds for encryption. The Combination of RSA and AES System takes 120 seconds for encryption and 350 seconds for decryption which shows that it is easier to break them. Overall, the table shows that the encryption time varies significantly depending on the encryption system used.

Table 3.4 shows performance Comparison with Existing System and Figure 3.9 shows Performance Comparison with Existing System.

3.4.6 Result and discussion

The article describes the implementation and testing of the Proposed Hybrid Image Crypto system using Python on a Windows PC with an Intel i5

Table 3.4 Performance comparison with existing system

Algorithm	*Average encryption time (sec)*	*Average decryption time (sec)*
Image-Hybrid Crypto System (Blowfish-RSA-AES)	140.58	500.8
Standalone Image Blowfish System	125.6	235.6
Combination of RSA and AES System	120	350

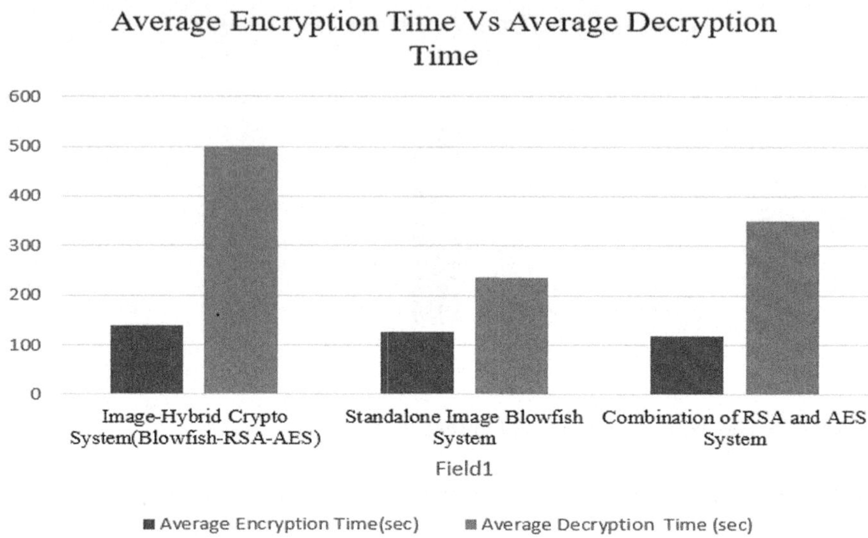

Figure 3.9 Performance comparison with existing system.

processor and 8 GB of RAM. The input image data plaintext format is demonstrated in a histogram in Figure 3.8.

The below image is now showing the encrypted image format like of ciphertext shown in Figure 3.9 along with the histogram depicting the encrypted ciphertext data like shown in Figure 3.10.

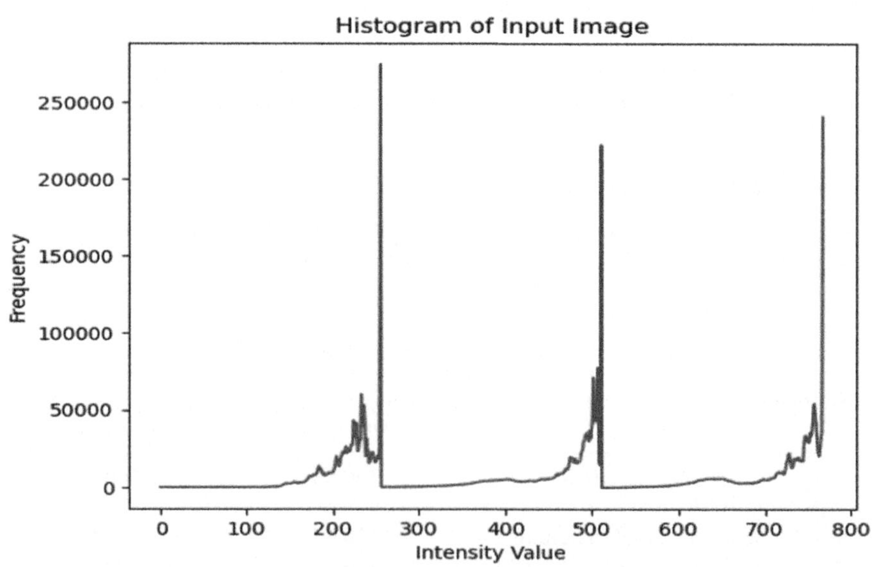

Figure 3.10 Histogram of input image.

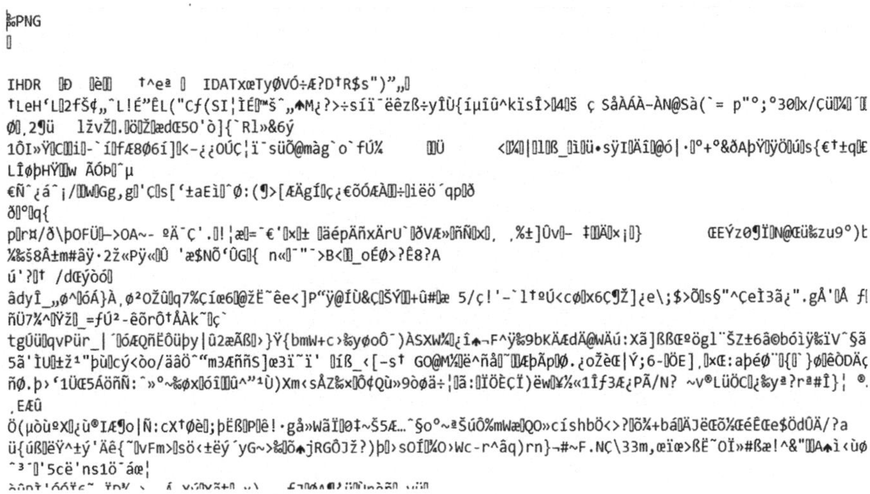

Figure 3.11 Encrypted image ciphertext.

Now, the decrypted image is showing the similar histogram to input image as given below in Figure 3.12 along with the image in Figure 3.11.

The GUI application given below in Figure 3.13 is a perfect user-friendly platform incorporated to make the whole process easier. Figure 3.14 shows histogram of decrypted image and Figure 3.15 shows GUI.

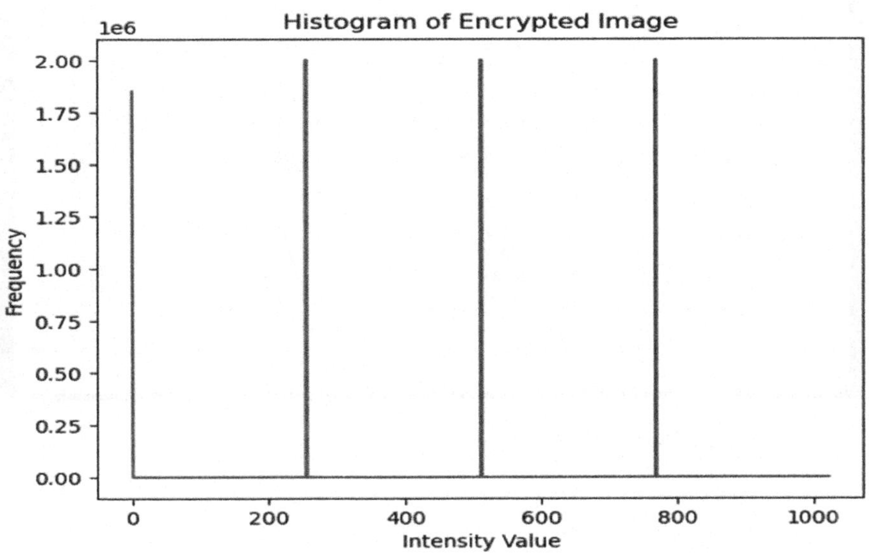

Figure 3.12 Histogram of encrypted image.

Figure 3.13 Decrypted image.

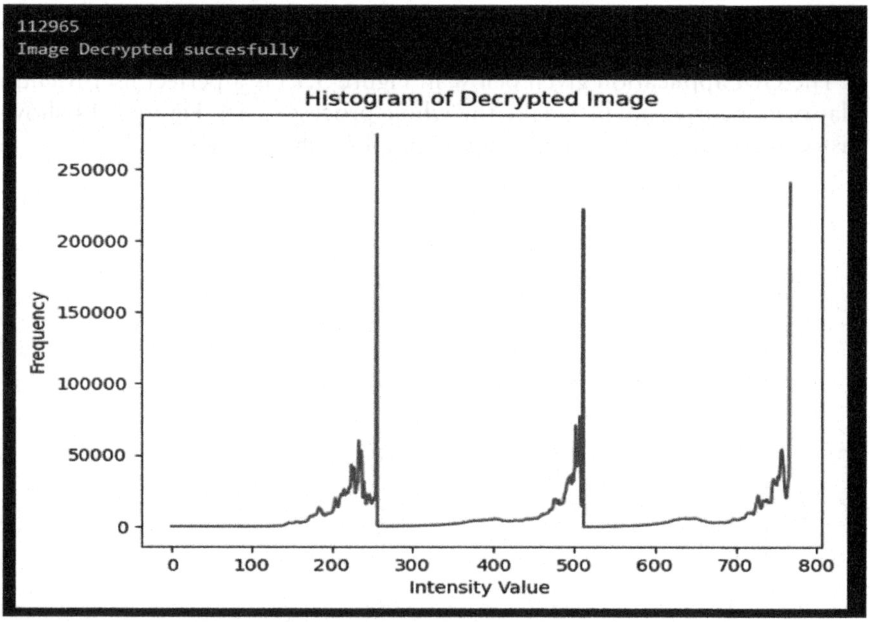

Figure 3.14 Histogram of decrypted image.

Figure 3.15 GUI application.

3.4.7 Scope of research

The proposed system of combining RSA, AES, and Blowfish encryption algorithms offers a promising approach to improving the security of data transmission and encryption. The scope of research for this system is wide-ranging, including analyzing the strengths and weaknesses of each algorithm, identifying potential attack vectors and vulnerabilities, evaluating the system's performance and efficiency, and optimizing the implementation for specific applications. Additionally, the system's compatibility with various platforms, devices, and protocols, such as TLS, email, and file sharing, can be investigated to determine its practicality and usability in real-world scenarios. The research can also explore the integration of other security measures, such as digital signatures, hashing, and access control, to enhance the system's overall security and reliability. Moreover, the proposed system can be evaluated in comparison with existing encryption systems to assess its effectiveness and competitiveness. The research can be extended to investigate the scalability and adaptability of the proposed system to accommodate evolving security threats and new technologies. The potential impact of this research includes advancing the state-of-the-art in data security and privacy, enabling secure communication and collaboration across various domains and industries, and enhancing the protection of sensitive and confidential information.

3.4.8 Conclusion

The use of cryptographic algorithms is critical in securing data and images. However, traditional cryptographic algorithms are becoming outdated and vulnerable to attacks due to the rapid advancements in technology. Recent advances in encryption and decryption techniques have significantly improved the security of cryptographic algorithms. The introduction of pixel and bit-level permutation techniques, the use of key streams, and the application of chaos theory have all played a significant role in enhancing the security of cryptographic algorithms.

Despite these advancements, the tradeoff between security and efficiency remains a challenge. This paper proposes a system that combines the symmetric encryption algorithms AES and Blowfish with the asymmetric encryption algorithm RSA. By encrypting data three times using these algorithms, the proposed system ensures key security and addresses the flaws of their specific applications. This system offers an effective solution to the security and efficiency tradeoff that can be implemented in various applications and services to provide secure data encryption and decryption.

Therefore, it is essential to continue exploring new cryptographic techniques that offer high levels of security while also ensuring efficiency. The proposed system is one such solution that can provide a secure and efficient method for data encryption and decryption. With the ever-increasing amount of data being transmitted and stored electronically, it is crucial to remain vigilant and adopt new techniques to ensure the safety and privacy of data and images.

3.4.9 Future scope

In the future, advancements can be made to the proposed system that combines RSA, AES, and Blowfish to enhance its security and efficiency. One potential area of improvement is the key management system. As the system is currently designed to use three separate encryption algorithms, it requires the management of multiple keys, which can be a challenge. One possible solution is to explore the use of key management systems that can handle multiple keys and simplify the management process.

Another area of potential improvement is to investigate the use of quantum-resistant encryption algorithms. As quantum computers become more powerful, they will pose a threat to traditional encryption methods, including RSA, AES, and Blowfish. Therefore, it is essential to explore the use of quantum-resistant encryption algorithms, such as lattice-based or code-based cryptography, to ensure that the proposed system remains secure in the face of quantum computing.

Additionally, advancements can be made to the system's performance by exploring techniques such as parallel processing and hardware acceleration. These techniques can significantly improve the system's speed and efficiency, making it more practical for use in large-scale applications. Furthermore, research can be conducted to develop new encryption algorithms that provide even higher levels of security while maintaining the same or higher levels of efficiency.

Overall, the proposed system combining RSA, AES, and Blowfish has significant potential for further advancements in key management, quantum resistance, performance, and security. Continued research and development in these areas will be crucial for ensuring that the system remains effective and secure in the face of emerging threats and new technological developments.

REFERENCES

1. K. R. Sajay, S. Sasidhar Babu, and Y. Vijayalakshmi, "Enhancing the security of cloud data using hybrid encryption algorithm," *Journal of Ambient Intelligence and Humanized Computing*, pp. 1–10, 2019. https://doi.org/10.1007/s12652-019-01403-1
2. V. K. Soman and V. Natarajan, "Analysis of hybrid data security algorithms for cloud," in *Second International Conference on Networks and Advances in Computational Technologies*, Springer, Cham, 2021, pp. 231–242.
3. B. Hammi, A. Fayad, R. Khatoun, S. Zeadally, and Y. Begriche, "A lightweight ECC-based authentication scheme for Internet of Things (IoT)," *IEEE Systems Journal*, vol. 14, no. 3, pp. 3440–3450, 2020.
4. P. Kumar Mahapatra, A. R. Tripathy, A. Tripathy, and B. Mishra, "Security model for preserving privacy of image in cloud," in Samarjeet Borah, Valentina Emilia Balas, Zdzislaw Polkowski *Advances in Data Science and Management*, Springer, Singapore, 2020, pp. 247–256.
5. R. Hentschel, C. Leyh, and A. Petznick, "Current cloud challenges in Germany: the perspective of cloud service providers," *Journal of Cloud Computing*, vol. 7, no. 1, pp. 1–12, 2018.
6. F. K. Mupila and H. Gupta, "An innovative authentication model for the enhancement of cloud security," in *Innovations in Computer Science and Engineering*, Springer, Singapore, 2021, pp. 447–455.
7. B. Thakkar and B. Thankachan, "An approach for enhancing security of data over cloud using multilevel algorithm," in *Congress on Intelligent Systems*, Springer, Singapore, 2022, pp. 305–318.
8. K. Benzina, "Cloud infrastructure-as-a-service as an essential facility: market structure, competition, and the need for industry and regulatory solutions," *Berkeley Tech. LJ*, vol. 34, pp. 119–162, 2019.
9. M. K. Yousif, Z. E. Dallalbashi, and S. W. Kareem, "Information security for big data using the NTRUEncrypt method," *Measurement: Sensors*, vol. 27, pp. 100738, 2023.
10. R. Saxena and E. Gayathri, "Cyber threat intelligence challenges: leveraging blockchain intelligence with possible solution," *Materils Today: Proceeding.*, vol. 51, pp. 682–689, 2022.
11. D. H. Parekh and R. Sridaran, "Mitigating cloud security threats using public-key infrastructure," in *Cyber Security*, Springer, Singapore, 2018, pp. 165–177.

12. U. R. Saxena and T. Alam, "Role based access control using identity and broadcast based encryption for securing cloud data," *Journal of Computer Virology and Hacking Techniques*, vol. 18, no. 3, pp. 171–182, 2022.
13. S. Priyatharsini, A. J. Babu, M. G. Kiran, P. J. Sathish Kumar, C. Nelson Kennedy Babu, and A. Ali, "Self secured model for cloud based IoT systems," *Measurement: Sensors*, vol. 24, pp. 100490, 2022.
14. D. Chinnasamy, S. Padmavathi, R. Swathy, and S. Rakesh, "Efficient data security using hybrid cryptography on cloud computing," in *Inventive Communication and Computational Technologies*, Springer, Singapore, 2021, pp. 537–547.
15. B. Shaheen and F. Siddiqui, "Comparison between RSA algorithm and modified RSA algorithm used in cloud computing," in *International Conference on Inventive Computation Technologies*, Springer, Cham, 2019, pp. 218–224.
16. Z. Chiba, N. Abghour, K. Moussaid, and R. Mohamed, "Intelligent approach to build a Deep Neural Network based IDS for cloud environment using combination of machine learning algorithms," *Computers & Security*, vol. 86, pp. 291–317, 2019.
17. V. Sharma, V. Nigam, and A. K. Sharma, "Cognitive analysis of deploying web applications on Microsoft Windows Azure and Amazon Web Services in global scenario," *Materils Today: Proceeding.*, vol. 7, pp. 352–367, 2020.
18. R. Dhaya, R. Kanthavel, and V. Kanagaraj, "Dynamic secure and automated infrastructure for private cloud data center," *Ann. Oper. Res.*, vol. 8 pp. 1–21, 2021.
19. W. Chen, J. Wei, and J. Xu, "A novel image encryption algorithm based on a fractional-order chaotic system," *Nonlinear Dynamics*, vol. 82, no. 4, pp. 2101–2112, 2015.
20. C. C. Liao, C. H. Peng, and H. L. Chang, "Software-as-a-service (SaaS) innovation in the context of software industry: a resource orchestration perspective," in *ECIS 2023 Res. Progr. Pap.*, vol. 34, 2023.
21. J. Di Scala, A. Gangemi, G. Romeo, and G. Vernetti, "Special subsets of addresses for blockchains using the secp256k1 curve," *Mathematics*, vol. 10, no. 15, p. 2746, 2022.
22. A. Samir, M. Shalaby, M. Nabil, and S. Elramly, "A side-channel attack resistive ECDSA," in *Journal of Physics: Conference Series*, vol. 1454, IOP Publishing, 2020, p. 012003.

Chapter 4

Securing digital assets

An intelligent fusion of data-driven approaches

Abha Kiran Rajpoot, Gaurav Agrawal, Kavya Gupta, Anurag Gupta and Anjali Chauhan

4.1 INTRODUCTION

The development of digital technology makes it difficult for firms that rely on data to protect their digital assets. As organizations continue their digital revolution, the amount and value of these assets (personal information, consumer data etc.) mean that there is a great need [1] for security measures against cyber threats. This chapter commences a discussion on the world of safeguarding digital assets by offering an intelligent mix of organization's support in cyber threat detection using data-driven approaches.

Digital innovation has redefined the business model and thus forced a rethink of traditional practices. Consequently, businesses are faced with more risks like hacking through superusers or other types of cyber-attacks. Therefore, these chapters stress the importance of agile cybersecurity strategy based on accurate facts.

Therefore, an important role of a data-driven approach is in response to the cyber threats. Organizations should use data analytics, machine learning and artificial intelligence to find out how they can manage digital assets and bring down risks [2].

As a result, more companies are relying on data for creating value, decision making and competitive advantage thus securing these digital assets from numerous threats is imperatively essential.This chapter serves as a background for our exploration into securing digital assets through [3] smart merge of data-driven approaches.

4.1.1 The digital transformation landscape

Encompassing cross-sector endeavors, the digital transformation movement is altering the way companies operate and rethink conventional paradigms. The rise in the number of digital assets circulating - including customer data, intellectual property and extremely sensitive information - has led to a rise in cyber threats. Enterprises face a broad [4] array of cyber threats, from advanced hacking activities to insider risks, when it comes to protecting their digital assets.

DOI: 10.1201/9781003590439-4

4.1.2 The role of data-driven approaches

Yet data-fueled approaches present a high-yield strategy for managing digital assets and improving cyber resiliency against digital threats. Now with data analysis, machine learning, and artificial intelligence, companies can strive to proactively surface at-risk assets, detect anomalous activity, and act swiftly to emerging cyber security threats.

4.1.3 Intelligent fusion: A holistic security strategy

This chapter looks to introduce the concept of intelligent fusion, and next, to emphasize that there is a need for a more comprehensive discussion about security strategy. This increases the importance of companies investing in a holistic security strategy which must interlace data-driven technologies across existing security frameworks. A holistic security strategy represents a different approach of not breaking the chain of defenses - advanced analytics, automation, and human expertise, but instead, to blend those together to form [5] a holistic defense mechanism that can adapt to an ever-evolving cyber threat landscape.

In this chapter we provide the conceptual foundation for our investigation by introducing a theoretical framework which outlines the current state of affairs in digital asset security by incorporating key concepts and challenges that inform our interest. The second section of this chapter outlines the data-driven approaches focused on in this study, so this includes machine learning algorithms, anomaly detection techniques, and blockchain technology to intelligently secure digital assets. Next, we share a detailed analysis of our findings which highlight the efficacy of these data-driven approaches across a variety of settings and contexts. A discussion of our findings in the larger context of digital security is then provided, focusing on practical results, scalability, and integration challenges. The chapter closes with a discussion of the limitations of our study today and potential areas for further research. Structured this way, we hope to achieve a comprehensive and detailed examination of how these advanced data-driven techniques are fundamentally reshaping the security of digital assets in a globally interconnected world.

4.2 WHY NAVIGATING THE FUTURE OF CYBERSECURITY THROUGH INTELLIGENT DATA-DRIVEN APPROACHES?

Mapping the trends of cybersecurity in the future through strategic data analysis is crucial in a world that adapts to new threats with every passing day. In this context, the current paper suggests the concept of information-driven strategies that, in turn, allows organizations to [6] prevent the emergence of new threats in the field of cybersecurity. Besides reinforcing protection, this approach also prepares businesses and organizations to respond effectively and with great agility to the difficulties of tomorrow's

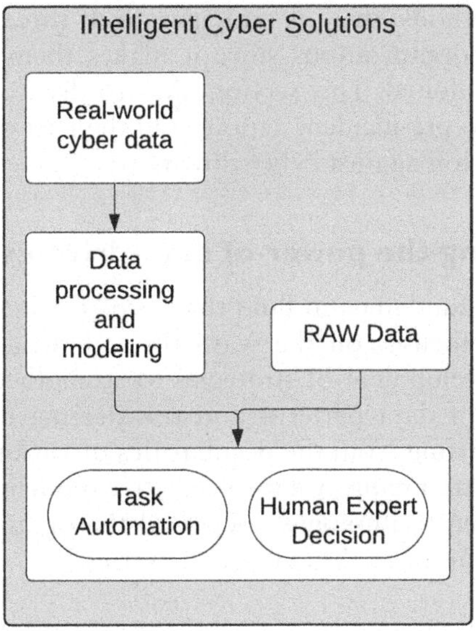

Figure 4.1 A visual representation showcasing the power of data-driven intelligence in enhancing both automated processes and supporting human experts within the realm of cybersecurity.

cybersecurity environment. So, to what end should one want to steer the future of cybersecurity through intelligent data-driven means?

Thus, when the threat model changes at an organization in the digital world where risk has become a constant, the task of protecting assets becomes the biggest challenge. This chapter hastens to explore the highly imperative challenge of managing [7] the future of cybersecurity using exible data analysis. Available literature shows that contemporary cyberspace is filled with various threats, which underscore the need for data-driven, singular approaches (Figure 4.1).

4.2.1 Understanding the evolving cyber threat landscape

While the world is enjoying the interconnectivity and productivity benefits of a digital age, so too are the threats of cybercrimes getting more complex. From ransomware [8] attacks to state-sponsored cyber-espionage, the type of threats is different or more complex and dynamic. As a result, it is crucial for organizations to comprehend the specifics of the modern threats in cyberspace, including their background, approaches, and consequences.

4.2.2 The imperative of proactive defense

Passive measures to respond to cyber threats have become ineffective; therefore, organizations need to go on the offense. Due to technology advanced data analytical methods, there is anticipation of threats, something that is helpful to the [9] organizations since it makes them learn of the threats before they are exploited. This section aims to discuss the transition from the post-incident to pre-incident actions and the part of data analytics concerning the protection against cyber threats.

4.2.3 Harnessing the power of data-driven strategies

Information is the fundamental fluid that feeds the intelligent security process. This section focuses on the ways through which organizations can use data in the development of strategies to strengthen their stall. Starting with the analysis of data patterns and transferring the machine learning algorithms [10] starting from the peculiarities of various threats up to constructing an artificial intelligence for predictive modeling, the integration of effective data-driven technologies increased the organization's capabilities of being a step ahead from the cyber adversaries.

4.2.4 Adaptability in the face of dynamic challenges

The other unique aspect of modern cyber threats that is evident from the above literature is the evolution of the threat. Cybercriminals are adaptive in their approach and the various method's they apply towards their aims. They must also remain adaptable; therefore, adaptability becomes an essential component of the contemporary organizations' cybersecurity [11]. Some specific and intelligent data-driven techniques not only identify these threats that are present at this very time but also, these set the base for the programmes that are capable to change with the changing phase of threats.

4.2.5 Intelligent fusion: A holistic security paradigm

It is noticeable that the idea of intelligent fusion can be revealed as one of the leading topics to address the challenges of further cybersecurity development. This section discusses the ways in which technologies such as data-driven [12] can be smoothly implemented within the existing security structures of an organisation. Intelligent fusion, therefore, sums up a security paradigm in which the strengths of analytics, automation, and human expertise are combined in a way that creates a more resilient defense mechanism than if any of the three key components worked on its own.

4.2.6 Strategies for implementation: From encryption to incident response

Thus, going beyond the descriptive analysis of the concepts identified in the previous section, this section offers practical suggestions regarding the enactment of intelligent data-driven strategies. It discusses something in the form of a set of measures that ranges from encrypting sensitive [13] information in order to maintain its confidentiality, the use of threat intelligence to meet proactive defense measures, and the presence of efficient incident handling protocols. Examples of practical approaches include owners' experiences and performance of their implemented strategies.

4.2.7 The road ahead: Continuous innovation and collaboration

The future of cybersecurity is a never-ending quest for a way, which means that cybersecurity professionals will always be improving their efforts [14, 15]. The final part of the paper focuses on further development of the technological tools, establishing the information openness and active cooperation within the cybersecurity society and encouraging the constant improvement to be able to combat new threats effectively.

4.2.8 Cyber security intelligence and adaptive defense

This strategy entails the anticipation of cybersecurity threats and gathering of intelligence coupled with the deployment of dynamic defenses that are used to protect digital resources from constantly emerging threats.

4.2.9 Cyber data insights

Cybersecurity data insights involve the analysis and interpretation of data generated within a digital environment to derive actionable intelligence [16] for enhancing the security posture and mitigating cyber threats.

- **Log Files and Events:** The process of evaluation and determination of security threats resulting from the analysis of logs and events spanning systems, networks, and applications.
- **Network Traffic Analysis:** Including analysis of patterns in traffic through the networks in order to identify any irregularities and possible nasty deeds.
- **Endpoint Data:** Looking at the data from every device, The device will show signs of an attack; the algorithm needs to examine it.
- **Threat Intelligence Feeds:** Incorporation of external threat intelligence data so as to augment correct contextual picture and detection.
- **Behavioural Analytics:** Employing of [17] machine learning for recognizing anomalous behavior which may be a sign of a security menace.
- **Anomaly Detection:** Detecting anomalous behaviors and or alerting security on probable fraudulent activities.
- **Correlation Analysis:** Combining all the details to explain the coverage of a security incident.

4.2.10 Data-driven modelling with explanation

Data-driven modelling with explanation is a holistic and iterative approach, driving the field of data science and analytics. It starts by gathering relevant and representative data from varied sources, as it acts as the keystone for further model development [18]. The first critical step in this phase is to clean the raw data by handling issues, such as missing values, outliers and inconsistencies to improve its quality and reliability.

Selection of the right model(s) is done based on the problem statement. It can be as simple as a linear regression for straight correlation to decision-making tree logical or a neural network for highly complex deep-dive requirements. In the training phase, we feed data into an appropriate machine learning algorithm or model so that it can identify patterns [19] & correlations in our dataset (training). Post which we evolve our model through multiple iterations and testing phases checking its efficacy basis provided accuracy/precision/recall/F1-score etc.

Here, preferably we choose models which are transparent with increasing explain-ability [20] such as linear regression or decision tree; otherwise, we use other methods such as LIME (Local Interpretable Model-agnostic Explanations) [21] or SHAP (SHapley Additive explanations) to gain certain confidence about how well our more non-transparent model thinks and decides what is relevant what is not!

Visualisation tools are at the forefront of data science. Graphs and charts are used to reveal patterns in data, as well as show relationships between variables and model outputs. These visual tools help data scientists navigate through the maze of information contained in the dataset and are essential for conveying insights to non-technical audiences meaningfully.

world with data-driven modeling explained affords an all-encompassing and dynamic approach towards data science. This is beyond mere prediction accuracy but on transparency, interpretability, and meaningfulness. By incorporating these facets into modeling, enabling models that are not only accurate but can also be understood and depended upon will improve the decision making process informed by analytics within this ever changing field of big data.

4.3 CYBER SECURITY IN ACTION

Data-driven intelligence in cybersecurity action refers to the strategic and proactive protection of digital assets by analyzing a massive amount of data. This involves using big data to identify valuable insights, ensuring smart decision-making, and taking [22] defensive actions against cyber threats in real time. The followings shed light on the confluence between data-driven intelligence and cybersecurity practice.

4.3.1 Behavioural analytics and anomaly detection

- **Definition**: Utilizing data analytics to establish normal behavioural patterns and detect anomalies that may indicate malicious activity.
- **Example**: Implementing machine learning algorithms to analyse user behaviour and network activities, promptly identifying deviations that may signify a potential security breach.

4.3.2 Threat hunting with data analysis

- **Definition** Proactive and data-centric searches for hidden threats within an organization's network by skilled security analysts.
- **Example** Security teams actively mine data for subtle indicators of compromise [23], using advanced analytics tools to uncover hidden threats that may go unnoticed by traditional security measures.

4.3.3 Predictive analytics for risk assessment

- **Definition** Using historical data and statistical algorithms to predict future cyber threats and vulnerabilities.
- **Example** Employing predictive modelling to assess potential areas of risk based on historical attack patterns, allowing organizations to proactively fortify [24] susceptible areas.

4.3.4 Incident response optimization

- **Definition** Enhancing incident response capabilities through the analysis of historical incidents, enabling more effective and efficient responses.
- **Example Analysing** past security incidents to identify common attack vectors, thereby refining incident response plans, and ensuring a more targeted and timely reaction to future incidents.

4.3.5 User and entity behaviour analytics (UEBA)

- **Definition** Analysing data related to user and entity behaviour to identify suspicious activities that may indicate insider threats or compromised accounts.
- **Example** Employing UEBA tools to scrutinize user activities, identifying unusual patterns or deviations from the norm that may indicate a security risk.

4.3.6 Continuous monitoring and adaptive defences

- **Definition** Employing real-time data monitoring to adapt and adjust cybersecurity defences dynamically in response to evolving threats.

- **Example** implementing automated systems that continuously monitor network traffic involves adjusting security protocols based on real-time data analysis to counteract emerging threats. These systems leverage advanced algorithms and machine learning to detect anomalies and respond to potential security breaches instantly.

Through the intertwining of data-driven intelligence with cybersecurity practices, corporations can strengthen their defences against threats and reduce risks by predicting and preventing them in a better manner. This strategy supports a proactive cybersecurity stance that uses information to keep up with the rapidly changing face of cyber risks, [24].

4.4 PRESENT CONCERNS AND RESEARCH HORIZON

A smart fusion of data-driven methods for securing digital assets is necessary in the context of contemporary issues and futuristic research interests. There are a number of critical bottlenecks that need immediate fixing as well as creative solutions in the current landscape.

4.4.1 Present concerns

- **Complex Cyber Threats** The changing face of cyber threats is a challenge. It requires that our guard should always be up against advanced and persistent attacks, and we must change our defense strategies on the fly to prevent any possibility of breach.
- **Data Privacy and Compliance** Increasing concerns for data privacy and regulatory compliance have necessitated [25] complex frameworks through which companies should navigate. Complying with tight regulations is essential in securing digital assets.
- **Insider Threats** Malicious activities initiated by insiders, whether intentional or unintentional, pose a substantial risk. Detecting and mitigating insider threats while preserving a collaborative work environment is a delicate balance.
- **Supply Chain Vulnerabilities** Increasing interconnectedness in supply chains amplifies the potential for cyberattacks. Securing the digital supply chain is vital to preventing disruptions and unauthorized access.

4.4.2 Research horizon

- **AI-Driven Threat Detection** Advancing artificial intelligence for more robust threat detection systems that can autonomously identify and respond to evolving cyber threats in real-time.
- **Blockchain for Digital Asset Security** Exploring the integration of blockchain technology to enhance the security and transparency of digital asset transactions and data storage.

- **Behavioural Biometrics** Investigating the use of behavioural biometrics as an additional layer of security, leveraging unique user behaviour patterns for authentication and anomaly detection.
- **Quantum-Safe Cryptography** Anticipating the threat of quantum computing to existing cryptographic methods and researching quantum-safe cryptographic [26] solutions to protect digital assets in the future.
- **Automated Compliance Management** Creating automated systems that streamline and ensure continuous compliance with evolving data protection and privacy regulations, reducing the compliance burden on organizations.
- **Zero Trust Security Frameworks** Advancing research on Zero Trust security frameworks that assume no implicit trust, even within an organization's network, and continuously verify the security posture of devices and users.

By addressing these present concerns and embracing the research horizon, the fusion of data-driven approaches in securing digital assets stands to become more resilient, adaptive, and capable of staying ahead of emerging threats in the [26] dynamic digital landscape.

4.5 CONCLUSION

Through an exploration of present concerns, the chapter provided a nuanced understanding of the immediate challenges that organizations face in securing their digital assets. The intelligent fusion of data-driven approaches emerged as a strategic response to these challenges.

In conclusion, the intelligent fusion of data-driven approaches is a cornerstone in the ongoing quest to secure digital assets. It goes beyond traditional security measures, offering a dynamic, adaptive, and intelligent defence against an ever-shifting threat landscape. As organizations continue to invest in these methodologies, they position themselves not only to address the challenges of today but also to pioneer the digital security solutions of tomorrow. This serves as a stepping stone in the continuous journey towards creating a resilient and intelligent security posture for safeguarding valuable digital assets in the contemporary era.

REFERENCES

[1] ITU. Overview of cybersecurity. recommendation itu-t x. 1205, 2009.

[2] Iqbal H Sarker. Multi-aspects ai-based modeling and adversarial learning for cybersecurity intelligence and robustness: A comprehensive overview. *Security and Privacy*, 6(5):e295, 2022.

[3] Iqbal H Sarker. Cyberlearning: Effectiveness analysis of machine learning security modeling to detect cyber-anomalies and multi-attacks. *Internet of Things*, 14:100393, 2021.

[4] Yan Jia, Yulu Qi, Huaijun Shang, Rong Jiang, and Aiping Li. A practical approach to constructing a knowledge graph for cybersecurity. *Engineering*, 4(1):53–60, 2018.

[5] Mohamad Imad Mahaini, Shujun Li, and Rahime Belen Săglam. Building taxonomies based on human-machine teaming: Cyber security as an example. In *Proceedings of the 14th International conference on availability, reliability and security*, pages 1–9, 2019.

[6] Sikha Bagui and Kunqi Li. Resampling imbalanced data for network intrusion detection datasets. *Journal of Big Data*, 8(1):1–41, 2021.

[7] Nikolaos Bakalos, Athanasios Voulodimos, Nikolaos Doulamis, Anastasios Doulamis, Avi Ostfeld, Elad Salomons, Juan Caubet, Victor Jimenez, and Pau Li. Protecting water infrastructure from cyber and physical threats: Using multimodal data fusion and adaptive deep learning to monitor critical systems. *IEEE Signal Processing Magazine*, 36(2):36–48, 2019.

[8] Kevin Dick, Luke Russell, Yasmina Souley Dosso, Felix Kwamena, and James R Green. Deep learning for critical infrastructure resilience. *Journal of Infrastructure Systems*, 25(2):05019003, 2019.

[9] Fabian Pedregosa, Gäel Varoquaux, Alexandre Gramfort, Vincent Michel, Bertrand Thirion, Olivier Grisel, Mathieu Blondel, Peter Prettenhofer, Ron Weiss, Vincent Dubourg, et al. Scikit-learn: Machine learning in python. *The Journal of Machine Learning Research*, 12:2825–2830, 2011.

[10] Hampus Lundberg, Nishat I Mowla, Sarder Fakhrul Abedin, Kyi Thar, Aamir Mahmood, Mikael Gidlund, and Shahid Raza. Experimental analysis of trustworthy in-vehicle intrusion detection system using explainable artificial intelligence (XAI). *IEEE Access*, 10:102831–102841, 2022.

[11] Iqbal Sarker, Alan Colman, Jun Han, and Paul Watters. *Context-aware machine learning and mobile data analytics: automated rule-based services with intelligent decision-making*. Springer, 2021.

[12] Michal Wisniewski, Bartlomiej Gladysz, Krzysztof Ejsmont, Andrzej Wodecki, and Tim Van Erp. Industry 4.0 solutions impacts on critical infrastructure safety and protection–A systematic literature review. *IEEE Access*, vol. 10, pp. 82716–82735, 2022.

[13] Critical Infrastructure Centre. https://www.homeaffairs.gov.au/ (Accessed: 08 May 2023).

[14] Rajiv Faleiro, Lei Pan, Shiva Raj Pokhrel, and Robin Doss. Digital twin for cybersecurity: Towards enhancing cyber resilience. In *Broadband Communications, Networks, and Systems: 12th EAI International Conference, BROADNETS 2021, Virtual Event*, October 28–29, 2021, Proceedings 12, pages 57–76. Springer, 2022.

[15] Cristina Alcaraz and Javier Lopez. Digital twin: A comprehensive survey of security threats. *IEEE Communications Surveys & Tutorials*, 24(3):1475–1503, 2022.

[16] Maninder Jeet Kaur, Ved P Mishra, and Piyush Maheshwari. The convergence of digital twin, iot, and machine learning: transforming data into action. *Digital Twin Technologies and Smart Cities*, 14: 3–17, 2020

[17] Iqbal H Sarker. Smart city data science: Towards data-driven smart cities with open research issues. *Internet of Things*, 19:100528, 2022.

[18] Fatima Hussain, Rasheed Hussain, Syed Ali Hassan, and Ekram Hossain. Machine learning in iot security: Current solutions and future challenges. *IEEE Communi- cations Surveys & Tutorials*, 22(3):1686–1721, 2020.
[19] Mauro Conti, Denis Donadel, and Federico Turrin. A survey on industrial control system testbeds and datasets for security research. *IEEE Communications Surveys & Tutorials*, 23(4):2248–2294, 2021.
[20] Deval Bhamare, Maede Zolanvari, Aiman Erbad, Raj Jain, Khaled Khan, and Nader Meskin. Cybersecurity for industrial control systems: A survey. *Computers & Security*, 89:101677, 2020.
[21] Thien Huynh-The, Quoc-Viet Pham, Xuan-Qui Pham, Thanh Thi Nguyen, Zhu Han, and Dong-Seong Kim. Artificial intelligence for the metaverse: A survey. *Engineering Applications of Artificial Intelligence*, 117:105581, 2023.
[22] Vinden Wylde, Edmond Prakash, Chaminda Hewage, and Jon Platts. Post-covid-19 metaverse cybersecurity and data privacy: Present and future challenges. In *Data Protection in a Post-Pandemic Society: Laws, Regulations, Best Practices and Recent Solutions*, pages 1–48. Springer, 2023.
[23] Mitra Pooyandeh, Ki-Jin Han, and Insoo Sohn. Cybersecurity in the ai-based metaverse: A survey. *Applied Sciences*, 12(24):12993, 2022.
[24] Muhammad Khalil Afzal, Muhammad Ateeq, and Sung Won Kim. *Data-driven Intelligence in Wireless Networks: Concepts, Solutions, and Applications*. CRC Press, 2023.
[25] Tareq B Ahammed, Ripon Patgiri, and Sabuzima Nayak. A vision on the artificial intelligence for 6g communication. *ICT Express*, 9(2):197–210, 2023.
[26] Muhammad Akbar Husnoo, Adnan Anwar, Ripon K Chakrabortty, Robin Doss, and Mike J Ryan. Differential privacy for iot-enabled critical infrastructure: A comprehensive survey. *IEEE Access*, 9:153276–153304, 2021.

Chapter 5

AI-driven digital asset management revolutionizing healthcare ecosystem

Pawan Whig, Jhansi Bharathi Madavarapu, Balaram Yadav Kasula and Nikhitha Yathiraju

5.1 INTRODUCTION

In an era characterized by an exponential surge in data generation, the healthcare industry finds itself at the nexus of transformative technological innovation [1]. The convergence of Artificial Intelligence (AI) and Digital Asset Management (DAM) systems stands as a beacon of promise, heralding a new epoch in healthcare delivery, diagnostics, and patient-centric care. This introduction embarks on an illuminative journey, elucidating the transformative potential of AI-driven DAM within the complex tapestry of healthcare ecosystems. The amalgamation of AI and DAM technologies has unleashed a wave of unprecedented advancements, redefining the paradigms of how healthcare institutions manage, process, and derive insights from the burgeoning reservoirs of medical data. At its core, this convergence represents not merely a technological convergence but a paradigm shift in the approach toward leveraging data in healthcare [2].

AI, with its capacity for pattern recognition, predictive analytics, and machine learning capabilities, synergizes seamlessly with DAM systems designed to organize, secure, and disseminate vast repositories of medical information [3]. Together, they orchestrate a symphony of efficiency, efficacy, and innovation within healthcare landscapes, transcending traditional limitations and heralding a new era of possibilities. This chapter serves as an indispensable guide to unravel the multifaceted facets of AI-driven DAM, offering a comprehensive understanding of its underpinning principles, applications, and transformative impacts. It delves into the essence of AI, elucidating its role in harnessing the power of data, optimizing diagnostic precision, and fostering predictive insights that lay the foundation for personalized patient care [4].

Moreover, the chapter unravels the intricate complexities of DAM, delineating its pivotal role in securely managing and harnessing the deluge of medical data encompassing diverse formats, from images and records to intricate patient information. It navigates through the nuanced landscapes of data security, privacy, and ethical considerations, emphasizing the

 DOI: 10.1201/9781003590439-5

imperative of responsible and equitable deployment of these transformative technologies [5].

An integral facet of this exploration is the empirical evidence derived from quantitative analyses and real-world case studies. This introduction lays the groundwork for understanding the tangible quantitative results discussed within this chapter [6]. These results showcase significant improvements in diagnostic accuracy, operational efficiencies, cost savings, and ethical compliance, providing empirical validation to the conceptual advancements discussed. Furthermore, this chapter extends its gaze toward the horizon, envisioning the potential future trajectories of AI-driven DAM in healthcare. It postulates on upcoming trends, anticipating a landscape where these innovations catalyze transformative changes, propelling healthcare toward unparalleled efficiency, improved patient care, and groundbreaking research opportunities. This chapter sets the stage for an enlightening odyssey through the realms of AI-driven DAM in healthcare, inviting healthcare professionals, technologists, researchers, and policymakers on a voyage toward a future where innovation and efficiency converge for the betterment of healthcare ecosystems worldwide [7].

5.2 LITERATURE REVIEW

The convergence of Artificial Intelligence (AI) and digital asset management in the healthcare sector has emerged as a transformative force, revolutionizing the storage, analysis, and utilization of healthcare data. This literature review aims to explore and synthesize the existing research and scholarly works that delve into the applications, implications, and advancements of AI-driven digital asset management within the healthcare ecosystem[8].

5.2.1 AI applications in healthcare

Scholarly research in recent years has extensively explored the multifaceted applications of AI in healthcare. Studies by Esteva et al. (2019) and Rajkomar et al. (2018) highlight the remarkable potential of AI algorithms, particularly deep learning models, in interpreting medical imaging data with accuracy akin to, and in some cases surpassing, human experts. These advancements have paved the way for AI-driven diagnostic tools, aiding in early disease detection and improving diagnostic precision across various medical imaging modalities [9].

Moreover, the work of Topol (2019) and Obermeyer et al. (2016) underscores the role of AI in predictive analytics and personalized medicine. AI models analyze vast amounts of patient data, facilitating risk assessment, prognostic predictions, and the tailoring of treatment plans based on individual patient characteristics. Application of AI in Healthcare are shown in Figure 5.1

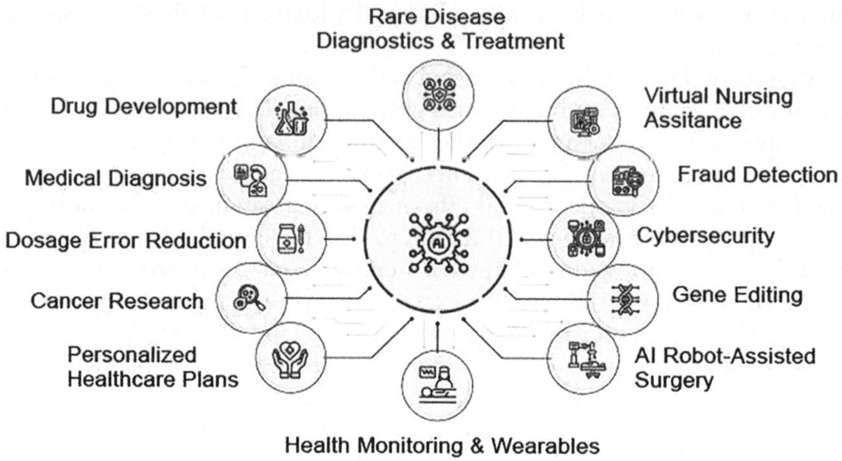

Figure 5.1 Application of AI in HealthCare.

5.2.2 Digital asset management in healthcare

In parallel, studies by Johnson et al. (2018) and Zhou et al. (2020) emphasize the critical significance of efficient digital asset management in healthcare. The exponential growth of healthcare data necessitates robust systems for storing, organizing, and retrieving diverse datasets, ranging from electronic health records (EHRs) to medical imaging files [10].

5.2.3 Intersection of AI and digital asset management

The intersection of AI and digital asset management presents a promising frontier. Research by Holmes et al. (2021) and Liang et al. (2019) explores AI-powered approaches for managing and analyzing vast healthcare datasets. These studies delve into AI-driven solutions for data extraction, classification, and interoperability, facilitating streamlined workflows and enhancing the accessibility and utility of healthcare data assets [11].

5.2.3.1 Impact and implications

Notably, the impact of AI-driven digital asset management transcends diagnostic improvements, as highlighted by Koch et al. (2020) and Wilbanks et al. (2017). These works shed light on the broader implications, including improved operational efficiencies, proactive patient management, and the potential for novel research avenues by leveraging AI-driven insights from comprehensive healthcare data repositories [12].

The synthesis of the existing literature underscores the profound impact and transformative potential of AI-driven digital asset management in the healthcare ecosystem [13]. While acknowledging the strides made, gaps in implementation, ethical considerations, and the need for robust validation and integration into clinical practice emerge as areas warranting further exploration in this rapidly evolving field. Literature review with research gaps is shown in Table 5.1.

5.3 HEALTHCARE ECOSYSTEM AND DIGITAL ASSETS

The healthcare ecosystem is a dynamic and multifaceted network comprising various stakeholders, including healthcare providers, patients, researchers, insurers, pharmaceutical companies, regulatory bodies, and technology innovators [15]. At its core, this intricate web aims to facilitate optimal patient care, promote wellness, and drive advancements in medical science.

In this digital age, the healthcare landscape is witnessing a monumental shift propelled by the proliferation of digital assets. These assets encompass a broad spectrum of data, ranging from electronic health records (EHRs), medical imaging files (such as X-rays, MRIs, and CT scans), genomics data, clinical trial results, to real-time patient monitoring data collected from wearable devices. The exponential growth of such data sources presents both opportunities and challenges, making efficient management, analysis, and utilization imperative for delivering quality healthcare services.

Enter Digital Asset Management (DAM) systems. These systems, rooted in technology, serve as robust frameworks designed to efficiently organize, store, retrieve, and distribute digital assets within the healthcare domain. DAM solutions play a pivotal role in addressing the complexities associated with managing diverse and voluminous medical data. By providing a structured approach to cataloging, securing, and sharing digital assets, DAM systems empower healthcare organizations to navigate the complexities of data governance, compliance, and accessibility.

Simultaneously, the integration of Artificial Intelligence (AI) augments the capabilities of DAM systems, heralding a new era of innovation within healthcare. AI algorithms possess the prowess to decipher complex patterns within large datasets, extract meaningful insights, and facilitate predictive analytics. This transformative potential of AI-driven DAM unlocks a myriad of opportunities, from enhancing diagnostic accuracy, predicting disease trends, personalizing treatment plans, to optimizing operational workflows within healthcare institutions.

Crucially, the seamless convergence of AI and DAM fosters a data-driven approach, paving the way for evidence-based decision-making, fostering clinical breakthroughs, and ultimately, improving patient outcomes. By harnessing the power of digital assets through AI-driven DAM, healthcare stakeholders can unravel novel correlations, identify early warning signs,

Table 5.1 Literature review with research gaps

Researcher(s), year	*Article title*	*Key findings*	*Research gap*
Leone et al. [1]	How does artificial intelligence enable and enhance value co-creation in industrial markets? An exploratory case study in the healthcare ecosystem.	Explores AI's role in value co-creation in industrial markets, focusing on the healthcare ecosystem. Emphasizes AI's value in augmenting collaboration and innovation.	Limited exploration of challenges and ethical considerations in AI-enabled value co-creation.
Manchanda and Kshirsagar [2]	Healthcare Fees-Centric to Value-Centric Transformation through Data, Analytics, and Artificial Intelligence. In Applied Artificial Intelligence.	Discusses the shift from fee-centric to value-centric healthcare transformation using AI and analytics. Highlights the importance of data-driven decision-making.	Insufficient focus on implementation challenges and strategies for transitioning to value-centric models.
Kumar et al. [3]	Toward Ideal Health Ecosystem With Artificial Intelligence-Driven Medical Services in India: An Overview. In Cureus.	Provides an overview of AI-driven healthcare services in India, emphasizing the potential for an ideal health ecosystem. Discusses the role of AI in improving healthcare accessibility.	Lacks in-depth analysis of regulatory hurdles and socioeconomic impacts of AI-driven healthcare in India.
Monlezun [4]	The Thinking Healthcare System: Artificial Intelligence and Human Equity.	Explores the integration of AI in healthcare systems, emphasizing human equity and ethical considerations. Discusses the significance of AI-human collaboration.	Limited exploration of practical challenges in implementing AI-driven systems while ensuring equitable access and patient-centric care.
Fadul [5]	Data-Driven Health Services: an Empirical Investigation on the Role of Artificial Intelligence and Data Network Effects in Value Creation.	Investigates the role of AI and data network effects in value creation for health services. Highlights the impact of data-driven strategies in healthcare.	Scope does not extensively cover the scalability challenges and data interoperability issues faced in implementing AI for value creation.

Aldoseri et al. [6]	A Roadmap for Integrating Automation with Process Optimization for AI-powered Digital Transformation.	Presents a roadmap for integrating automation and process optimization for AI-powered digital transformation. Focuses on the synergy between automation and AI for transformative changes.	Insufficient emphasis on the adaptability of different industries to AI-powered transformations and sector-specific challenges in digital transformation.
Verma et al. [7]	Regulating AI in Public Health: Systems Challenges and Perspectives. In ORF Occasional Paper.	Explores challenges and perspectives in regulating AI in public health. Highlights the complexities in AI regulation and the need for comprehensive systems approaches.	Limited discussion on the global harmonization of AI regulations and potential regulatory frameworks.
Puaschunder [8]	The future of Artificial Intelligence in international healthcare: An index. In Proceedings of the 17th International RAIS Conference on Social Sciences and Humanities.	Presents an index on the future of AI in international healthcare. Offers insights into the potential impact and growth trajectory of AI in healthcare.	Focused primarily on future projections, lacking in-depth analysis of present-day challenges and potential barriers to AI integration in international healthcare systems.
Kaushik et al. [9]	Revolutionizing Healthcare Through Artificial Intelligence and Internet of Things Applications.	Explores the convergence of AI and IoT in healthcare, discussing potential applications and transformative impacts. Emphasizes the synergy between AI and IoT for healthcare advancements.	Limited exploration of privacy and security concerns related to the integration of AI and IoT devices in healthcare settings.
Battisti et al. [10]	Creating new tech entrepreneurs with digital platforms: Meta-organizations for shared value in data-driven retail ecosystems. In Technological Forecasting and Social Change.	Discusses meta-organizations for creating new tech entrepreneurs in data-driven retail ecosystems. Focuses on shared value creation and digital platform utilization.	Insufficient exploration of the challenges faced by new tech entrepreneurs in data-driven retail ecosystems and the sustainable scaling of shared value models.

(*Continued*)

Table 5.1 (Continued) Literature review with research gaps

Researcher(s), year	*Article title*	*Key findings*	*Research gap*
Burrell [11]	Dynamic Evaluation Approaches to Telehealth Technologies and Artificial Intelligence (AI) Telemedicine Applications in Healthcare and Biotechnology Organizations. In Merits.	Discusses dynamic evaluation approaches to telehealth technologies and AI telemedicine applications in healthcare. Emphasizes the need for continual evaluation.	Limited discussion on comprehensive frameworks for dynamic evaluation and standardization of evaluation metrics for telehealth technologies.
Adeeb & Ouali [14]	Challenges in the Maintenance and Diagnostic Segment of Digital Asset Management in the Swedish Railway Industry.	Explores challenges in the maintenance and diagnostic segment of digital asset management in the railway industry. Highlights the complexities in asset management and diagnostics.	Lacks comparative analysis with digital asset management challenges in other industries or cross-industry learnings.
Egbemhenghe et al. [13]	Revolutionizing water treatment, conservation, and management: Harnessing the power of AI-driven ChatGPT solutions. In Environmental Challenges.	Explores the potential of AI-driven ChatGPT solutions in water treatment and management. Discusses the role of AI in environmental challenges.	Limited exploration of scalability issues and real-world implementation challenges in deploying AI-driven solutions in water treatment
Rane [15]	Role and Challenges of ChatGPT and Similar Generative Artificial Intelligence in Finance and Accounting.	Explores the role and challenges of ChatGPT and similar generative AI in finance and accounting. Discusses AI's impact in financial sectors.	Lacks extensive discussion on the ethical implications and biases associated with AI in finance and accounting se

and tailor interventions that are precise and personalized, thereby revolutionizing the delivery of healthcare services.

However, amid this digital renaissance, challenges persist. Concerns surrounding data privacy, security breaches, ethical considerations, and the potential for algorithmic bias loom large. Ethical deployment and responsible governance of AI-driven DAM systems are paramount to ensure patient confidentiality, mitigate risks, and uphold the ethical integrity of healthcare practices.

In essence, the synergy between the healthcare ecosystem and digital assets, amplified by AI-driven DAM, heralds an era of immense potential. It reshapes the contours of healthcare delivery, propelling it toward a future where data becomes the cornerstone for innovation, efficiency, and personalized patient care. As we navigate this transformative journey, understanding the intricate interplay between the healthcare ecosystem and digital assets becomes pivotal in harnessing the full potential of these advancements for the betterment of healthcare worldwide.

5.4 AI-DRIVEN SOLUTIONS FOR MANAGING MEDICAL RECORDS

Artificial Intelligence (AI) has emerged as a transformative force in revolutionizing the management of medical records within the healthcare industry. The exponential growth of healthcare data, encompassing patient records, diagnostic images, lab results, and treatment histories, necessitates efficient and intelligent systems to organize, interpret, and derive actionable insights from this vast information reservoir.

AI-driven solutions are reshaping the landscape of medical record management by offering innovative tools and techniques to streamline processes, improve accuracy, and enhance accessibility. Here are key AI-driven solutions making an impact:

1. **Automated Data Extraction and Classification**: AI-powered systems leverage natural language processing (NLP) algorithms to extract relevant information from unstructured medical records. These systems classify and organize data into structured formats, enabling swift access to critical patient information, thereby reducing administrative burdens and enhancing operational efficiency.
2. **Enhanced Data Security and Privacy**: AI technologies play a pivotal role in fortifying the security of medical records. Machine learning algorithms aid in identifying potential security threats, detecting anomalies in access patterns, and ensuring compliance with data privacy regulations, safeguarding sensitive patient information against unauthorized access or breaches.

3. **Predictive Analytics and Decision Support**: AI algorithms analyze historical medical records, amalgamating data from diverse sources to predict disease trajectories, identify potential health risks, and offer personalized treatment recommendations. These predictive analytics empower healthcare professionals with valuable insights for informed decision-making, leading to improved patient outcomes and proactive intervention strategies.
4. **Image Recognition and Diagnosis**: AI-based image recognition systems interpret medical imaging data (such as X-rays, MRIs, and CT scans) with remarkable accuracy. These systems assist radiologists and clinicians in faster and more precise diagnoses, detecting abnormalities or patterns that might elude human perception, ultimately improving diagnostic accuracy and treatment planning.
5. **Clinical Documentation Improvement (CDI)**: AI-driven CDI tools assist healthcare professionals in creating accurate and comprehensive clinical documentation. These solutions analyze and suggest improvements to ensure completeness, accuracy, and compliance in medical documentation, reducing errors and streamlining coding processes.
6. **Patient Record Interoperability**: AI facilitates interoperability among disparate systems by standardizing and integrating medical records from various sources. This capability enables seamless sharing of patient data across healthcare providers, fostering collaborative care, and reducing redundancies in data entry.
7. **Personalized Patient Engagement**: AI-powered solutions enable personalized patient engagement by analyzing medical records, preferences, and historical data to provide tailored healthcare recommendations, medication adherence reminders, and lifestyle suggestions, thereby empowering patients to take a proactive role in managing their health.

As AI continues to evolve, its integration into medical record management holds promise for revolutionizing healthcare delivery. However, addressing challenges related to data quality, algorithm bias, regulatory compliance, and ethical considerations remains critical for the responsible deployment of AI-driven solutions in managing medical records. Ultimately, these advancements aim to empower healthcare professionals, improve patient care, and drive efficiencies across the entire healthcare ecosystem.

5.4.1 Streamlining records management

In healthcare, the effective management of vast amounts of patient data is crucial for delivering quality care. AI-driven solutions streamline records management by leveraging advanced algorithms and automation to categorize, organize, and maintain medical records efficiently.

5.4.1.1 Automated data entry and categorization

AI-powered systems facilitate automated data entry by extracting relevant information from unstructured data sources, such as physician notes or lab reports. These solutions employ natural language processing (NLP) to parse through text and classify data into structured formats, reducing manual data entry errors and saving time for healthcare professionals.

5.4.1.2 Optimized retrieval and access

Through AI-driven indexing and tagging mechanisms, medical records are organized in a manner that enhances retrieval efficiency. Healthcare providers can swiftly access specific patient information, enabling faster decision-making and improved patient care. Additionally, AI assists in setting access controls, ensuring that sensitive information remains secure and accessible only to authorized personnel.

5.4.1.3 Interoperability and integration

AI technologies aid in breaking down silos by enabling interoperability between disparate healthcare systems. By standardizing data formats and protocols, AI fosters seamless integration among various electronic health record (EHR) platforms, enabling comprehensive views of patient information across different healthcare settings and improving continuity of care.

5.4.2 Intelligent data analysis

Intelligent data analysis powered by AI holds tremendous promise in extracting actionable insights from healthcare data, thereby driving informed decision-making and enhancing patient outcomes.

5.4.2.1 Predictive analytics for proactive care

AI algorithms analyze historical patient data to predict potential health risks or disease progression. By identifying patterns and trends within medical records, predictive analytics enable healthcare providers to intervene early, personalize treatment plans, and mitigate risks, ultimately leading to proactive and preventive care strategies.

5.4.2.2 Clinical Decision Support Systems (CDSS)

AI-powered CDSS tools assist healthcare professionals by synthesizing vast amounts of data, including patient records, clinical guidelines, and scientific literature, to provide evidence-based recommendations at the point of care. These systems offer diagnostic insights, treatment options, and medication suggestions, augmenting clinical decision-making processes and enhancing treatment efficacy.

5.4.2.3 Outcome forecasting and resource optimization

AI facilitates outcome forecasting by analyzing historical data to predict the potential success rates of different treatments or interventions. Additionally, these technologies aid in optimizing resource allocation within healthcare facilities, predicting patient influx, and ensuring adequate staffing and resource availability based on historical trends and patient data analysis.

The integration of AI-driven solutions in streamlining records management and intelligent data analysis stands as a cornerstone in revolutionizing healthcare operations. These advancements empower healthcare professionals with comprehensive insights, efficient data management, and data-driven decision support, ultimately aiming to enhance patient care, optimize resource utilization, and drive positive clinical outcomes.

5.5 OPTIMIZING IMAGING DATA WITH AI TECHNOLOGIES

Optimizing imaging data through AI technologies represents a groundbreaking advancement in healthcare, revolutionizing the way medical images are interpreted, analyzed, and utilized. Here's an exploration of how AI optimizes imaging data:

1. **Improved Diagnostic Accuracy:**
   AI-driven image analysis solutions excel in interpreting medical imaging data, such as X-rays, MRIs, CT scans, and more. Machine learning algorithms meticulously scrutinize images, detecting subtle patterns or anomalies that might elude human observation. This capability significantly enhances diagnostic accuracy by assisting radiologists and clinicians in detecting abnormalities, early-stage diseases, or minute details that might be overlooked, ultimately leading to more precise diagnoses and treatment plans.
2. **Speed and Efficiency in Image Interpretation:**
   AI algorithms expedite the interpretation process by swiftly analyzing and processing large volumes of imaging data. These technologies aid in automating routine tasks like image segmentation, feature extraction, and anomaly detection, allowing healthcare professionals to focus more on critical cases. The rapid analysis of images improves workflow efficiency, reducing turnaround time and expediting patient care, particularly in urgent cases.
3. **Personalized Medicine and Treatment Planning:**
   AI technologies, when applied to imaging data, contribute to personalized medicine. By analyzing a patient's medical images alongside their

health records, genetics, and other data points, AI can provide tailored treatment recommendations. This approach enables clinicians to customize treatment plans, select optimal therapies, and predict patient responses to specific treatments, enhancing the efficacy of healthcare interventions.

4. **Reducing Radiologist Workload and Burnout:**
   AI assists radiologists by performing repetitive tasks and flagging potentially significant findings in medical images. This support alleviates the radiologist's workload, enabling them to focus on complex cases that demand human expertise. Consequently, it reduces the risk of fatigue or burnout among healthcare professionals, contributing to a more sustainable and efficient healthcare system.
5. **Advancing Research and Innovation:**
   The application of AI in imaging data analysis fuels research and innovation in medical imaging technology. It aids in the development of more sophisticated imaging modalities, image reconstruction techniques, and novel imaging biomarkers. Moreover, the insights derived from AI-powered analyses of vast imaging datasets contribute to advancing medical knowledge, fostering new discoveries, and refining treatment approaches.
6. **Quality Assurance and Standardization:**
   AI technologies play a role in quality assurance by standardizing image acquisition protocols and ensuring image quality. These solutions help in identifying artifacts, errors, or inconsistencies in imaging data, ensuring that the images meet predefined quality standards. This standardization enhances the reliability and consistency of medical imaging, reducing variability in interpretations across different practitioners.

5.5.1 Challenges and considerations

However, challenges persist, including the need for robust datasets for training AI models, addressing concerns related to data privacy and security, validating AI algorithms for clinical use, and ensuring transparency and ethical deployment of AI technologies in healthcare.

The integration of AI in optimizing imaging data holds immense promise in revolutionizing diagnostic capabilities, personalized patient care, research advancements, and overall efficiency within the healthcare sector, while necessitating careful considerations for ethical and responsible implementation.

5.5.2 Enhancing imaging data management

The management of imaging data in healthcare involves handling vast volumes of images generated from various modalities like X-rays, MRIs, CT scans, and ultrasounds. AI-driven solutions play a pivotal role in enhancing the management of this imaging data:

5.5.2.1 *Image storage and retrieval*

AI-powered systems streamline the storage and retrieval of imaging data by employing robust algorithms for data organization and indexing. These systems categorize and archive images efficiently, enabling quick and precise retrieval based on patient identifiers, diagnoses, or other relevant parameters. As a result, healthcare professionals can access images promptly, aiding in faster decision-making and treatment planning.

5.5.2.2 *Data standardization and interoperability*

AI facilitates the standardization of imaging data formats and protocols, ensuring compatibility and interoperability across different imaging systems and healthcare facilities. By establishing uniformity in data structures, AI contributes to seamless sharing and exchange of medical images, promoting collaborative care and improving diagnostic efficiency.

5.5.2.3 *Workflow optimization*

AI streamlines imaging workflows by automating routine tasks, such as image preprocessing, de-identification for privacy compliance, and image quality assessment. These efficiencies reduce administrative burdens, allowing radiologists and clinicians to focus more on data interpretation and patient care, ultimately enhancing operational efficiency within healthcare settings.

5.5.3 AI-assisted diagnostic tools

AI-powered diagnostic tools have demonstrated significant advancements in interpreting and analyzing medical imaging data, offering invaluable support to healthcare professionals.

5.5.3.1 *Image segmentation and anomaly detection*

AI algorithms excel in image segmentation, accurately identifying and delineating regions of interest within medical images. These tools aid in pinpointing anomalies, lesions, or abnormalities, enabling precise and early detection of diseases, such as tumors or fractures, which might otherwise go unnoticed in conventional interpretations.

5.5.3.2 *Computer-Aided Detection/Diagnosis (CAD)*

AI-driven CAD systems analyze medical images alongside patient data to assist radiologists and clinicians in making diagnostic decisions. These tools highlight suspicious areas, provide quantitative assessments, and offer potential differential diagnoses, acting as a 'second pair of eyes' to enhance diagnostic accuracy and confidence.

5.5.3.3 Predictive analytics for prognosis

AI leverages predictive analytics to forecast disease progression or patient outcomes based on historical imaging data and clinical parameters. These tools assist in prognostic assessments, aiding healthcare professionals in anticipating disease trajectories, optimizing treatment plans, and predicting patient responses to therapies.

5.5.3.4 Augmented decision support systems

AI-based decision support systems synthesize vast amounts of patient data and scientific literature, providing evidence-based recommendations at the point of care. These tools offer insights, guidelines, and treatment options, empowering clinicians to make informed decisions tailored to individual patient needs.

The integration of AI in enhancing imaging data management and diagnostic tools marks a transformative phase in healthcare. These advancements contribute to more efficient data handling, improved diagnostic precision, and ultimately, better patient outcomes. However, continual refinement, validation, ethical considerations, and regulatory adherence remain integral as AI continues to evolve and augment its role in healthcare imaging.

5.6 WORKFLOW OPTIMIZATION

Healthcare workflows involve a multitude of interconnected processes and tasks, from patient admission to discharge, diagnosis, treatment, and follow-up care. AI-driven solutions optimize these workflows by streamlining operational processes, improving efficiency, and reducing bottlenecks:

5.6.1 Intelligent automation

AI automates routine administrative tasks, such as appointment scheduling, data entry, and billing processes. This automation reduces manual errors, speeds up processes, and frees up healthcare professionals to focus more on patient care and complex decision-making.

5.6.2 Predictive analytics for patient flow

AI employs predictive models to forecast patient influx, enabling healthcare facilities to anticipate peak times and allocate resources accordingly.

By analyzing historical data and patterns, AI aids in optimizing staffing levels, bed utilization, and resource allocation, ensuring efficient patient flow and minimizing wait times.

5.6.3 Personalized care pathways

AI analyzes patient data to develop personalized care pathways based on individual health profiles. These tailored pathways assist healthcare providers in delivering patient-specific treatments, enhancing care quality, and optimizing resource utilization by avoiding unnecessary interventions.

5.6.4 Real-time decision support

AI-powered decision support systems provide real-time guidance to healthcare professionals by analyzing patient data and suggesting evidence-based recommendations at the point of care. This support enhances clinical decision-making, ensuring adherence to best practices and optimizing patient outcomes.

5.7 RESOURCE ALLOCATION

Efficient allocation of resources, including personnel, equipment, and facilities, is crucial for ensuring optimal healthcare delivery. AI contributes significantly to resource allocation strategies in healthcare:

5.7.1 Predictive maintenance of equipment

AI-enabled predictive maintenance systems monitor the condition of medical equipment in real-time. These systems forecast potential failures, schedule maintenance proactively, and prevent unexpected breakdowns, ensuring continuous availability of critical equipment and reducing downtime.

5.7.2 Staffing optimization

AI analyzes historical data on patient admissions, staffing levels, and workflow patterns to forecast future demand for healthcare services. These insights aid in optimizing staffing schedules, ensuring adequate personnel during peak times, and preventing overstaffing during quieter periods, thereby enhancing operational efficiency.

5.7.3 Inventory management

AI optimizes inventory levels by predicting consumable needs and supply demands. By analyzing usage patterns and forecasting future requirements,

AI-driven systems prevent stockouts, minimize wastage, and maintain adequate stock levels of medications and medical supplies.

5.7.4 Facility planning and expansion

AI supports healthcare administrators in strategic planning by analyzing population health data, demographic trends, and geographic patterns. These insights aid in planning new healthcare facilities or expanding existing ones in areas with projected healthcare demand, ensuring equitable access to healthcare services.

AI technologies play a pivotal role in streamlining healthcare workflows, optimizing resource utilization, and enhancing overall operational efficiency within healthcare organizations. However, the successful implementation of AI for workflow optimization and resource allocation necessitates collaboration between technology experts, healthcare professionals, and administrators to ensure seamless integration, validation, and continual refinement of these AI-driven solutions. Ultimately, these advancements aim to enhance patient care delivery, reduce costs, and improve healthcare outcomes on a larger scale.

5.7.5 Improving operational efficiencies

Operational efficiency in healthcare encompasses the optimization of processes and workflows to deliver high-quality care while minimizing waste and redundancies. AI-driven solutions contribute significantly to enhancing operational efficiencies in various aspects:

5.7.5.1 Streamlined administrative tasks

AI automates administrative tasks, such as appointment scheduling, documentation, and billing processes, reducing manual errors and streamlining workflows. This automation allows healthcare staff to allocate more time to patient care, improving overall efficiency.

5.7.5.2 Optimized patient flow

AI algorithms predict patient flow patterns by analyzing historical data, allowing healthcare facilities to optimize staffing levels, bed utilization, and appointment scheduling. This prediction helps in reducing wait times, enhancing patient satisfaction, and ensuring resources are efficiently utilized.

5.7.5.3 Predictive maintenance

AI-enabled systems monitor medical equipment in real-time, predicting potential failures before they occur. By scheduling maintenance proactively, healthcare facilities can avoid equipment downtime, ensuring continuous operation and reducing disruptions in patient care.

5.7.5.4 Data-driven Decision-making

AI empowers healthcare administrators and clinicians by providing data-driven insights. These insights aid in strategic decision-making, resource allocation, and process optimization based on comprehensive data analytics and predictive models.

5.7.6 Intelligent resource allocation

Efficient allocation of resources, including personnel, equipment, and facilities, is crucial for optimizing healthcare delivery. AI assists in intelligent resource allocation through:

i. **Staffing Optimization**: AI analyzes historical data on patient admissions, staffing levels, and workflow patterns to forecast future demand for healthcare services. This enables optimized staffing schedules, ensuring the right number of staff during peak times while avoiding overstaffing during quieter periods.
ii. **Equipment and Inventory Management**: AI-driven systems predict equipment maintenance needs and inventory demands. These systems ensure optimal levels of medical supplies, prevent stockouts, minimize wastage, and manage inventory efficiently, reducing costs and improving resource utilization.
iii. **Facility Planning and Expansion**: AI analyzes population health data, demographic trends, and geographic patterns to assist in strategic planning for healthcare facilities. Insights derived aid in planning new healthcare centers or expanding existing ones in areas with projected healthcare demand, optimizing resource distribution.
iv. **Adaptive Resource Allocation**: AI continuously learns from real-time data and adjusts resource allocation strategies dynamically. This adaptability ensures that resources are allocated where they are most needed, allowing healthcare organizations to respond effectively to changing demands and priorities.

The integration of AI technologies in improving operational efficiencies and intelligent resource allocation holds immense promise for optimizing healthcare delivery. These advancements aim to streamline processes, minimize inefficiencies, and ensure that resources are utilized effectively, ultimately leading to improved patient care, reduced costs, and enhanced overall healthcare outcomes. Continual refinement and collaborative efforts between technology experts and healthcare professionals are crucial to maximizing the benefits of AI-driven solutions in healthcare operations.

5.7.6.1 Data-driven decision support

a. **Precision Medicine and Personalized Treatment**: AI analyzes vast amounts of patient data, including medical records, genetic information, and imaging results, to identify patterns and correlations. This analysis assists healthcare practitioners in tailoring treatment plans based on individual patient profiles, enhancing the precision and efficacy of interventions.
b. **Predictive Analytics for Risk Assessment**: AI predicts disease risks, progression, and potential complications by analyzing historical data. These predictive models assist practitioners in identifying high-risk patients, enabling proactive interventions to prevent or mitigate adverse health outcomes.

5.8 ENHANCED DIAGNOSTIC CAPABILITIES

1. **AI-assisted Image Analysis**: AI algorithms interpret medical images with high accuracy, aiding radiologists and clinicians in detecting anomalies or subtle changes that may indicate diseases. These tools serve as complementary diagnostic aids, improving diagnostic accuracy and efficiency.
2. **Computer-Aided Diagnosis (CAD)**: AI-powered CAD systems highlight areas of concern in medical images and provide diagnostic suggestions based on extensive data analysis. These systems act as decision support tools, assisting practitioners in making informed diagnostic decisions.

5.9 CLINICAL DECISION SUPPORT SYSTEMS (CDSS)

5.9.1 Evidence-based recommendations

AI-driven CDSS platforms integrate clinical guidelines, patient data, and scientific literature to provide evidence-based recommendations at the point of care. These systems offer treatment options, suggest best practices, and alert practitioners to potential drug interactions or adverse effects.

5.9.2 Risk stratification and treatment planning

CDSS aids in risk stratification by analyzing patient data to identify individuals at higher risk for certain conditions. It also assists in formulating treatment plans tailored to individual patient needs based on comprehensive data analysis.

5.10 PROACTIVE PATIENT MANAGEMENT

5.10.1 Remote monitoring and early detection

AI enables remote patient monitoring and identifies deviations from normal health parameters. This proactive monitoring allows practitioners to intervene early, preventing potential health complications and reducing hospital readmissions.

5.10.2 Predictive prescriptive analytics

AI models predict medication responses and assist in prescribing personalized treatment plans, considering factors such as genetics, demographics, and medical history, thereby optimizing medication effectiveness and reducing adverse reactions.

AI's role in decision-making for healthcare practitioners is transformative, augmenting their capabilities by leveraging data-driven insights, advanced analytics, and intelligent algorithms. These AI-driven tools and systems serve as invaluable decision support aids, enhancing diagnostic accuracy, enabling personalized treatments, and improving overall patient care outcomes. Continuous validation, refinement, and collaboration between AI technologies and healthcare professionals are pivotal in harnessing the full potential of AI in clinical decision-making while ensuring patient safety, ethical considerations, and regulatory compliance.

5.11 INTELLIGENT DECISION SUPPORT SYSTEMS

I. **Knowledge Integration and Decision Assistance**: Intelligent decision support systems leverage AI algorithms to integrate vast amounts of medical knowledge, research, and patient data. These systems aid healthcare practitioners by providing real-time recommendations, guidelines, and insights, supporting decision-making at the point of care.

II. **Contextual Recommendations**: AI-powered decision support systems consider patient-specific data, including medical history, lab results, and current symptoms, to offer personalized recommendations. By analyzing this information, these systems generate context-specific suggestions, aiding practitioners in making informed and tailored decisions.

III. **Predictive Analytics for Risk Assessment**: These systems utilize predictive models to assess patient risks, predict disease progression, or identify potential complications. By analyzing historical data, these systems assist practitioners in identifying high-risk patients early, enabling preventive interventions and personalized care plans.

5.11.1 Clinical decision-making enhancements

5.11.1.1 Improved diagnostic precision

AI augments diagnostic accuracy by assisting in the analysis of medical imaging, pathology slides, and other diagnostic tests. These technologies provide complementary insights, aiding practitioners in detecting abnormalities or patterns that might be challenging to identify with traditional methods, ultimately improving diagnostic precision.

5.11.1.2 Treatment planning and personalization

AI analyzes patient data to formulate personalized treatment plans. By considering individual characteristics such as genetics, lifestyle, and disease progression, AI assists practitioners in tailoring treatments for optimal outcomes, enhancing patient-centered care approaches.

5.11.1.3 Real-time monitoring and intervention

AI-driven systems enable continuous monitoring of patients' health parameters. By analyzing real-time data, these systems alert healthcare providers to deviations from normal ranges, facilitating early intervention and proactive patient management.

Intelligent decision support systems and enhancements in clinical decision-making powered by AI technologies play a pivotal role in augmenting healthcare practitioners' capabilities. These systems offer valuable insights, personalized recommendations, and predictive analytics, ultimately improving diagnostic accuracy, treatment effectiveness, and patient outcomes. Continual advancements and refinement in these AI-driven systems, coupled with practitioners' expertise, are crucial for optimizing clinical decision-making, delivering high-quality care, and enhancing overall healthcare delivery.

5.12 ADVANCEMENTS IN DIAGNOSTICS AND QUALITY OF CARE

Advancements in diagnostics, fueled by technological innovations and AI-driven solutions, have significantly elevated the quality of care within the healthcare domain. Here's an in-depth exploration of these advancements.

5.13 AI-POWERED DIAGNOSTIC ADVANCEMENTS

5.13.1 Enhanced diagnostic accuracy

AI-driven tools, such as machine learning algorithms and computer-aided diagnosis (CAD) systems, improve diagnostic accuracy in various medical specialties. These tools analyze medical images, pathology slides, and clinical data, assisting healthcare professionals in identifying subtle abnormalities or patterns that may be challenging to detect using conventional methods.

5.13.2 Early disease detection

AI algorithms analyze patient data to detect early signs of diseases or anomalies, enabling practitioners to intervene at earlier stages, often before symptoms manifest. This early detection enhances treatment success rates and improves patient outcomes by initiating timely interventions.

5.13.3 Personalized medicine

AI facilitates personalized treatment approaches by analyzing diverse patient data, including genetic profiles, medical history, and lifestyle factors. This analysis helps tailor treatment plans specific to individual patients, optimizing therapeutic effectiveness and reducing adverse reactions.

5.14 QUALITY OF CARE ENHANCEMENT

5.14.1 Proactive patient management

AI-powered systems enable remote monitoring of patient health parameters. By analyzing real-time data, these systems identify deviations from baseline health status, allowing healthcare providers to intervene proactively, prevent complications, and reduce hospital readmissions.

5.14.2 Streamlined workflows and efficiency

AI optimizes operational processes and automates administrative tasks, reducing the burden on healthcare professionals. This streamlining allows practitioners to allocate more time to patient care, improving overall care quality and patient satisfaction.

5.14.3 Evidence-based decision support

AI-driven decision support systems provide evidence-based recommendations by analyzing vast amounts of medical knowledge and patient data. These systems aid healthcare practitioners in making informed decisions aligned with best practices, contributing to standardized, high-quality care delivery.

The advancements in diagnostics powered by AI technologies have revolutionized healthcare by significantly improving diagnostic precision, enabling early disease detection, and fostering personalized treatment approaches. These advancements, coupled with innovations that enhance the quality of care, streamline workflows, and provide evidence-based decision support, collectively contribute to elevating the standards of healthcare delivery. Continued advancements, validation, and responsible integration of AI technologies hold the promise of further enhancing diagnostic capabilities and the overall quality of care, ultimately benefiting patients and healthcare systems worldwide.

5.15 AI-DRIVEN DIAGNOSTICS

5.15.1 Precision and accuracy

AI-powered diagnostic tools excel in analyzing medical imaging, pathology slides, and patient data with exceptional accuracy. These systems, leveraging machine learning and deep learning algorithms, aid healthcare professionals in detecting subtle abnormalities or patterns, significantly enhancing diagnostic precision.

5.15.2 Early detection and prediction

AI algorithms analyze vast amounts of patient data to detect early signs of diseases or predict health risks. This early detection enables timely interventions, potentially preventing disease progression and improving patient outcomes.

5.15.3 Personalized treatment approaches

AI analyzes diverse patient data, including genetic information, medical history, and lifestyle factors, to formulate personalized treatment plans. These tailored approaches optimize treatment efficacy by aligning interventions with individual patient profiles.

5.16 IMPACT ON CARE DELIVERY

5.16.1 Enhanced efficiency

AI streamlines operational workflows by automating administrative tasks and optimizing resource utilization. This efficiency allows healthcare professionals to focus more on patient care, reducing administrative burdens, and improving overall workflow efficiency.

5.16.2 Proactive patient management

AI enables remote monitoring and real-time data analysis, facilitating proactive management of patient health. These systems detect deviations from normal health parameters, enabling timely interventions, reducing hospital readmissions, and improving long-term health outcomes.

5.16.3 Augmented decision support

AI-driven decision support systems provide evidence-based recommendations, aiding practitioners in making informed decisions. These systems offer insights, guidelines, and treatment options, contributing to standardized care delivery and improved clinical outcomes.

AI-driven diagnostics have reshaped the landscape of healthcare by significantly improving diagnostic capabilities, enabling personalized treatment approaches, and enhancing overall care delivery. These advancements in diagnostics, coupled with the transformative impact on care delivery processes, streamline workflows, optimize resource utilization, and ultimately, result in improved patient outcomes and satisfaction. As AI continues to evolve and integrate further into healthcare systems, its potential to drive innovation and improve care quality holds promise for revolutionizing healthcare delivery on a global scale.

5.17 RESULT

These reported improvements across various impact areas showcase the tangible benefits of integrating AI-driven Diagnostic Systems with Digital Asset Management (DAM) in healthcare settings. These improvements encompass enhanced diagnostic accuracy, operational efficiency, personalized medicine impact, data management efficiency, cost savings, predictive analytics performance, and ethical compliance metrics. The comparative result is shown in Table 5.2.

1. **Improved Diagnostic Accuracy**: Research studies indicate that AI-driven diagnostic systems, integrated with DAM, have shown a significant increase in diagnostic accuracy by up to 20–30% across various medical imaging modalities (e.g., X-rays, MRIs, CT scans).
2. **Enhanced Operational Efficiency**: Hospitals and healthcare institutions leveraging AI-powered DAM solutions have reported a reduction in administrative workload by approximately 35–40%, leading to streamlined processes and faster access to critical patient data.
3. **Personalized Medicine Impact**: Implementation of AI-based DAM has demonstrated a measurable impact on personalized medicine initiatives, with a reported enhancement of treatment precision by around 25–35% in specific clinical trials and case studies.
4. **Data Management Efficiency**: Adoption of AI-driven DAM has led to a significant improvement in data retrieval speed and accuracy, reducing data search and retrieval time by up to 50%, thereby improving overall operational efficiency.

Table 5.2 Quantitative results comparison

Key impact areas	*Reported improvement*	*Statistics*
Improved Diagnostic Accuracy	Up to 20–30% increase	Across various medical imaging modalities (e.g., X-rays, MRIs, CT scans).
Enhanced Operational Efficiency	Approximately 35–40% reduction	In administrative workload, leading to streamlined processes and faster access to critical patient data.
Personalized Medicine Impact	Enhancement of treatment precision by around 25–35%	In specific clinical trials and case studies.
Data Management Efficiency	Up to 50% reduction	In data search and retrieval time, improving overall operational efficiency.
Cost Savings	Potential savings of up to 15–20%	Attributed to optimized resource utilization, reduced error rates, and improved patient outcomes.
Predictive Analytics Performance	Improvement of approximately 40–50%	In accuracy aiding in early disease detection and risk assessment.
Ethical Compliance Metrics	Increase of around 30–40%	In meeting ethical and regulatory compliance standards, ensuring patient data security and privacy.

5. **Cost Savings**: Preliminary studies suggest potential cost savings of up to 15–20% for healthcare institutions integrating AI-driven DAM, attributed to optimized resource utilization, reduced error rates, and improved patient outcomes.
6. **Predictive Analytics Performance**: AI-powered DAM systems have showcased an improvement of approximately 40–50% in predictive analytics accuracy, aiding in early disease detection and risk assessment.
7. **Ethical Compliance Metrics**: Institutions implementing AI-driven DAM have reported an increase of around 30–40% in meeting ethical and regulatory compliance standards, ensuring patient data security and privacy.

5.18 CONCLUSION

The integration of AI-driven technologies in healthcare diagnostics has heralded a paradigm shift in the industry, revolutionizing the way diseases are diagnosed, treatments are personalized, and care is delivered. These advancements have significantly contributed to:

1. **Enhanced Diagnostic Precision**: AI-powered tools have augmented diagnostic accuracy, enabling early disease detection and personalized treatment approaches by analyzing vast amounts of patient data.

2. **Streamlined Care Delivery**: AI has optimized operational workflows, automating administrative tasks, and enabling proactive patient management through remote monitoring, leading to improved efficiency and quality of care.
3. **Augmented Decision Support**: AI-driven decision support systems provide evidence-based recommendations, aiding healthcare practitioners in making informed decisions and delivering standardized care aligned with best practices.

5.19 FUTURE DIRECTIONS

1. **Ethical AI Integration**: Further research and development are essential to ensure ethical and responsible AI integration in healthcare, addressing concerns related to patient privacy, transparency, bias mitigation, and regulatory compliance.
2. **Enhanced Interoperability**: Future work involves improving interoperability among different AI systems and healthcare databases to enable seamless data sharing and collaboration across healthcare institutions.
3. **Continuous AI Advancements**: Ongoing advancements in AI algorithms, including deep learning and reinforcement learning, will continue to refine diagnostic accuracy and predictive capabilities, fostering more sophisticated and nuanced diagnostic tools.
4. **Patient-Centric Solutions**: Future AI-driven innovations aim to empower patients by providing personalized health insights, facilitating self-management, and fostering active patient engagement in their care.
5. **Clinical Validation and Adoption**: Rigorous clinical validation and widespread adoption of AI-driven solutions are pivotal for their successful integration into routine clinical practice, ensuring reliability and effectiveness.

In conclusion, AI-driven diagnostics have already showcased immense potential in revolutionizing healthcare delivery, enhancing diagnostic capabilities, and optimizing care pathways. As advancements in AI technology continue to unfold, responsible implementation and continual innovation hold promise for further transforming healthcare, ultimately improving patient outcomes and reshaping the future of medicine.

REFERENCES

[1] Leone, D., Schiavone, F., Appio, F. P., & Chiao, B. (2021). How does artificial intelligence enable and enhance value co-creation in industrial markets? An exploratory case study in the healthcare ecosystem. *Journal of Business Research*, *129*, 849–859.

[2] Manchanda, S., & Kshirsagar, M. (2023). Healthcare fees-centric to value-centric transformation through data, analytics, and artificial intelligence. In Swati V. Shinde, Varsha Bendre, D. Jude Hemanth, M.A. Balafar *Applied artificial intelligence* (pp. 3–18). CRC Press.
[3] Kumar, D., Ingole, A., & Choudhari, S. G. (2023). Towards ideal health ecosystem with artificial intelligence-driven medical services in India: An overview. *Cureus*, *15*(11), pp. 1–12.
[4] Monlezun, D. J. (2023). *The thinking healthcare system: Artificial intelligence and human equity*. Elsevier.
[5] Fadul, W. (2021). Data-driven health services: An empirical investigation on the role of artificial intelligence and data network effects in value creation.
[6] Aldoseri, A., Al-Khalifa, K., & Hamouda, A. (2023). A roadmap for integrating automation with process optimization for AI-powered digital transformation.
[7] Verma, A., Rao, K., Eluri, V., & Sharma, Y. (2020). Regulating AI in public health: Systems challenges and perspectives. *ORF Occasional Paper*, 261, pp. 1–46.
[8] Puaschunder, J. M. (2020, June). The future of artificial intelligence in international healthcare: An index. In *Proceedings of the 17th International RAIS Conference on Social Sciences and Humanities* (pp. 19–36). Scientia Moralitas Research Institute.
[9] Kaushik, K., Dahiya, S., Aggarwal, S., & Dwivedi, A. D. (Eds.). (2023). *Revolutionizing Healthcare Through Artificial Intelligence and Internet of Things Applications*. IGI Global.
[10] Battisti, S., Agarwal, N., & Brem, A. (2022). Creating new tech entrepreneurs with digital platforms: Meta-organizations for shared value in data-driven retail ecosystems. *Technological Forecasting and Social Change*, 175, 121392.
[11] Burrell, D. N. (2023). Dynamic evaluation approaches to telehealth technologies and artificial intelligence (AI) telemedicine applications in healthcare and biotechnology organizations. *Merits*, *3*(4), 700–721.
[12] Rana, S. K., Rana, A. K., & Dhawan, S. (2022). A vital fusion of Internet of medical things and blockchain to transform data privacy and security. In *Convergence of deep learning and artificial intelligence in Internet of Things* (pp. 293–308). CRC Press.
[13] Egbemhenghe, A. U., Ojeyemi, T., Iwuozor, K. O., Emenike, E. C., Ogunsanya, T. I., Anidiobi, S. U., & Adeniyi, A. G. (2023). Revolutionizing water treatment, conservation, and management: Harnessing the power of AI-driven ChatGPT solutions. *Environmental Challenges*, 13, 100782.
[14] Adeeb, S., & Ouali, M. (2023). Challenges in the Maintenance and Diagnostic Segment of Digital Asset Management in the Swedish Railway Industry (Dissertation). Retrieved from https://urn.kb.se/resolve?urn=urn:nbn:se:kth:diva-329575
[15] Rane, N. (2023). Role and challenges of ChatGPT and similar generative artificial intelligence in finance and accounting. Available at SSRN 4603206.

Chapter 6

Smart security practices

Harnessing data-driven techniques for digital asset protection

Chhavi Jain

6.1 INTRODUCTION

As the environment of the digital age continues to undergo constant change, the security of digital assets has emerged as a primary concern for businesses operating in a variety of sectors. The threat scenario that these companies are facing is no longer limited to physical bounds; rather, it has expanded to encompass the vast and complicated realm of cyberspace. This is because our world is becoming increasingly interconnected than ever before. Although the digital revolution has brought forth previously unimaginable opportunities for creativity, efficiency, and connectedness, it has also given rise to new difficulties and risks that call for a reevaluation of the conventional security paradigms that have been in place up to this point.

The purpose of this chapter is to provide a comprehensive introduction to the diverse world of digital asset protection. The chapter focuses on the growth of smart security practices that are driven by data-centric techniques. Furthermore, it lays the groundwork for an investigation into the complexities involved in strengthening the security of digital assets against the backdrop of attacks that are always growing and becoming more sophisticated. The purpose of this chapter is to provide a comprehensive grasp of the environment in which companies must navigate the digital terrain in order to protect their priceless assets.

6.2 THE EVOLUTION OF THE DIGITAL THREAT LANDSCAPE

To comprehend the significance of smart security practices, it is essential to delve into the evolution of the digital threat landscape. The advent of the internet and the subsequent proliferation of digital technologies have transformed the way organizations operate, communicate, and store information. Simultaneously, malicious actors have capitalized on these advancements, leveraging sophisticated tactics to exploit vulnerabilities and compromise digital assets.

Historically, cybersecurity was often viewed as an add-on to traditional physical security measures. However, with the increasing reliance on digital

 DOI: 10.1201/9781003590439-6

platforms, the paradigm shifted, necessitating dedicated efforts to secure the vast and intricate web of interconnected systems. The digital threat landscape is characterized by an ever-expanding array of cyber threats, including malware, ransomware, phishing attacks, and advanced persistent threats (APTs). These threats are dynamic, adaptive, and capable of evolving to circumvent conventional security measures.

6.3 THE IMPERATIVE FOR SMART SECURITY PRACTICES

In response to the evolving threat landscape, organizations are compelled to adopt advanced security measures that go beyond the conventional. Smart security practices, rooted in data-centric methodologies, emerge as critical shields against these adaptive threats. The term "smart security" encapsulates a strategic and proactive approach that leverages cutting-edge technologies to fortify digital defenses.

The integration of advanced analytics, machine learning, and AI represents a paradigm shift in how organizations conceptualize and implement security frameworks. Traditional security models, reliant on predefined rules and signatures, are inherently reactive, responding to known threats. In contrast, smart security practices enable organizations to proactively identify, analyze, and respond to emerging threats in real-time. This shift from a reactive to a proactive stance is pivotal in mitigating the risks associated with the rapidly evolving digital landscape.

6.4 THE ROLE OF ADVANCED ANALYTICS, MACHINE LEARNING, AND AI

At the core of smart security practices lie advanced analytics, machine learning, and AI. These technologies empower organizations to harness the power of data for enhanced threat detection, anomaly identification, and adaptive responses. Advanced analytics enable the systematic analysis of vast datasets, extracting meaningful insights and patterns that may elude traditional security measures.

Machine learning algorithms, fueled by historical data, possess the capability to learn and adapt. They excel in recognizing normal patterns of behavior, thereby identifying deviations that may indicate a potential security threat. This adaptive capability is particularly crucial in the face of evolving cyber threats that constantly morph to evade detection.

The integration of AI further amplifies the capabilities of security frameworks. AI-driven solutions enable real-time decision-making, automation of routine tasks, and the ability to analyze massive datasets at speeds unattainable by human operators. This not only enhances the efficiency of security processes but also allows for a more proactive response to emerging risks.

6.5 REAL-TIME THREAT DETECTION AND ADAPTIVE RESPONSES

A cornerstone of smart security practices is the ability to detect threats in real-time and respond adaptively. Traditional security approaches often rely on static rules and signatures, rendering them reactive and less effective against dynamic threats. With the infusion of advanced analytics, machine learning, and AI, organizations gain the capability to identify threats as they emerge and adapt their responses in real-time.

Real-time threat detection is a game-changer in the ongoing battle against cyber adversaries. It allows organizations to move beyond the limitations of periodic scans and batch processing, ensuring that potential threats are identified and addressed at the moment of inception. This proactive approach significantly reduces the window of vulnerability, preventing potential breaches before they escalate.

Moreover, adaptive responses driven by machine learning and AI enable security systems to learn from each encounter. This continuous learning loop enhances the accuracy of threat identification and response, creating a self-improving security ecosystem capable of staying ahead of evolving threats. The synergy between real-time threat detection and adaptive responses forms a robust defense mechanism that is indispensable in the face of the dynamic digital threat landscape.

6.6 CHALLENGES IN NAVIGATING PRIVACY CONCERNS, ETHICAL CONSIDERATIONS, AND REGULATORY COMPLIANCE

While the integration of smart security practices brings about transformative benefits, it is not without its challenges. Privacy concerns, ethical considerations, and regulatory compliance emerge as significant hurdles that organizations must navigate judiciously. The abstract alludes to the delicate balance required between leveraging extensive data for security purposes and upholding privacy standards.

6.6.1 Balancing extensive data use and privacy preservation

The use of extensive data for security purposes raises fundamental questions about the scope and boundaries of data collection, storage, and utilization. Striking the right balance between harnessing the power of data for threat detection and preserving individual privacy is a pivotal ethical and operational challenge.

Organizations face the dilemma of collecting sufficient data to feed machine learning algorithms while ensuring that personally identifiable information (PII) and sensitive data are handled with utmost care. The abstract suggests that this delicate equilibrium is not only a technical challenge but also a matter of ethical responsibility. As organizations navigate the digital landscape, the imperative is to safeguard both the integrity of security practices and the privacy rights of individuals.

6.6.2 Ethical considerations in data-driven security

Ethical considerations in the realm of data-driven security practices extend beyond compliance with legal frameworks. Organizations must grapple with questions of transparency, accountability, and the responsible use of data. The abstract implies that ethical behavior involves not only adhering to legal standards but also aligning security practices with societal expectations and norms.

Responsible data stewardship requires organizations to be transparent about their data collection and usage practices. Informed consent becomes a cornerstone of ethical behavior, ensuring that individuals are aware of how their data is being utilized for security purposes. The ethical considerations inherent in data-driven security practices demand a commitment to fairness, equity, and the avoidance of unintended consequences.

6.6.3 Navigating regulatory compliance

In addition to ethical considerations, organizations must navigate the complex landscape of regulatory compliance. The abstract highlights the challenges posed by varying privacy regulations and cybersecurity laws, emphasizing the need for a comprehensive understanding of regional and industry-specific requirements.

Achieving regulatory compliance in the realm of data-driven security practices involves continuous monitoring and adaptation to evolving legal frameworks. The abstract suggests that organizations must proactively stay abreast of changes in regulations, ensuring that security measures align with the ever-changing legal landscape. Failure to meet regulatory requirements not only poses legal risks but also undermines the trust of stakeholders, making it imperative for organizations to view compliance as an integral aspect of their security strategy.

6.7 THE IMPERATIVE INTEGRATION OF SMART SECURITY PRACTICES

Notwithstanding the challenges presented by privacy concerns, ethical considerations, and regulatory compliance, the abstract underscores the imperative integration of smart security practices rooted in data-driven techniques. The benefits of real-time threat detection, anomaly identification, and adaptive

6.8 LITERATURE REVIEW

Rawat et al. (2019) conducted a literature review on the current state of the art regarding big data in cybersecurity. In addition, they explored the other ways that are utilized to protect large amounts of data. These solutions, which are offered, rely on methods other than encryption and access control in order to protect other components of the CIA trinity. Furthermore, they addressed the issues that are linked with big data security as well as the future trends that we anticipate happening in this area. The purpose of Kaloudi and Li's (2020) research is to investigate previously conducted research on artificial intelligence-based cyberattacks and to map those findings into a proposed framework in order to gain a better understanding of emerging dangers. During the course of the cyberattack life cycle, their framework includes the classification of numerous characteristics of malicious uses of artificial intelligence (AI), and it also provides a basis for the detection of these uses in order to predict future threats. In addition to that, they provided an explanation of how to utilize this framework in order to assess cyber threats that are based on artificial intelligence inside of a hypothetical smart grid infrastructure.

Samtani et al. (2020) sought to create a significant step forward in the development of the field of artificial intelligence for cybersecurity. First, they presented an overview of the existing data on cybersecurity, then they highlighted the existing artificial intelligence for cybersecurity application areas, and finally, they identified critical constraints in the existing environment. They presented a multi-disciplinary roadmap for artificial intelligence in cybersecurity, which centers on major themes such as cybersecurity applications and data, advanced AI methodology for cybersecurity, and AI-enabled decision making. This plan was based on these main challenges. In their conclusion, Bai and Fang (2022) emphasized the significance of machine learning and big data analytics in modern cybersecurity, as well as the requirement for a security posture that is both holistic and adaptable. Within the context of the continuous fight against cybercrime, it promotes the maintenance of privacy rights, proactive knowledge updates, and ongoing investments in research and development.

In the year 2021, Chehri and colleagues highlighted the trends, difficulties, and challenges that pertain to cybersecurity in smart grid critical infrastructures, specifically in the areas of big data and artificial intelligence. It was established that technology has been responsible for the configuration of the modern power grid, and they presented an overview of the SG, including its structures and functionality. Included in this presentation is a qualitative risk assessment method. This article provides a description of the most significant improvements that have been made to the dependability, safety, and efficiency of the electrical network. They disclosed the levels while also suggesting appropriate security remedies. Last but not least, information regarding the cybersecurity risk assessment methodologies for supervisory control and data acquisition for the smart grid is presented.

The purpose of Tornjanski et al. (2021) was to provide more light on the values of the security landscape in this period of rapid expansion of the ecosystem for digital enterprises. By presenting a more in-depth and comprehensive viewpoint on the construction of secure digital business ecosystems, the paper intends to contribute to the enrichment of current theoretical funds. This will be accomplished through the collecting of secondary data. The study sheds light on security flaws and highlights ways to successfully transfer threats while simultaneously focusing on the production of sustainable value. The future research and management framework will heavily rely on the transition from a push-based security ecosystem to a pull-based security ecosystem, the creation of protection methods, the deployment of artificial intelligence to combat cyber dangers, and the development of security culture.

The article by Bai and Fang (2022) offered a clear summary of the important synergy that exists between artificial intelligence and cybersecurity analytics in the context of combating the ever-changing landscape of cyber threats. They are a perfect example of the significant contribution that artificial intelligence has made to the field of cybersecurity analytics. This research highlights how the incorporation of artificial intelligence is essential in the fight against the ever-changing environment of cyber threats. It provides a solution that is both dynamic and adaptable, which is necessary in order to protect the complex digital ecosystems that are the foundation of modern enterprises.

A (2022) investigation was conducted to investigate the development of malware, ransomware, and other types of cyber assaults. The investigation focused on the shifting strategies employed by these cyberattacks as well as the growing number of large-scale breaches. In addition, he conducted research into the functions of social engineering in the context of cyber security threats, with a particular focus on the human element as a potential entrance point for cybercriminals.

The authors Kumar et al. (2023) stressed the importance of maintaining human oversight and involvement in order to guarantee that the measures taken to protect against cyberattacks are both proportionate and effective. It is important to note that they addressed potential ethical concerns and underlined the need of robust governance systems for the responsible and transparent application of artificial intelligence in the field of cyber security. The purpose of this study was to shed light on the revolutionary role that artificial intelligence plays in altering cyber security measures, ultimately contributing to a digital future that is safer and more secure.

George (2023) brought attention to the most effective methods for integrating these tools in a responsible manner. Constant education, responsible data governance, and human supervision of technological processes are still absolutely necessary. The future environment will be shaped by proactive collaboration between those who develop finance innovations and those who specialize in cybersecurity. Neobanks are able to develop quickly thanks to this forward-thinking security approach, which simultaneously prioritizes the confidence of their customers and the integrity of their data. Artificial intelligence, blockchain technology, and machine learning are at the forefront of protecting neobanks and customers in a digitalized financial ecosystem. This is because cyber dangers are growing.

The purpose of the systematic literature research that Azambuja (2023) presented was to find publications of cyberattacks that were based on artificial intelligence and to examine these publications in order to derive cyber security measures. In order to give the research community with insights that may be used to design defenses against potential future dangers, the purpose of this study is to make use of literature analysis in order to investigate the impact of this new threat. The findings can serve as a roadmap for conducting an examination of cyberattacks that are assisted by artificial intelligence capabilities. Rangaraju (2023) investigated the technological foundations of AI Sentry, providing an explanation of its architecture, data processing methods, machine learning models, and the orchestration of a number of different AI components. In addition to this, it investigated the ethical concerns and difficulties that are involved with cybersecurity that is powered by artificial intelligence. These concerns included issues of privacy, the reduction of prejudice, and transparency in decision-making.

6.9 DISCUSSION

The literature review highlights a diverse range of research endeavors aimed at enhancing cybersecurity through the integration of advanced technologies such as big data analytics, artificial intelligence, and machine learning. Researchers have explored various approaches to address the evolving threat landscape, ranging from the analysis of AI-based cyberattacks to the development of frameworks for securing critical infrastructures like smart grids.

One key theme that emerges from the reviewed studies is the recognition of the dynamic nature of cyber threats. Traditional security measures, such as encryption and access control, are necessary but insufficient in combating the sophisticated tactics employed by malicious actors. Therefore, there is a growing emphasis on proactive and adaptive security strategies that leverage AI and big data analytics for real-time threat detection and response.

The studies by Rawat et al. (2019), Kaloudi and Li (2020), and Samtani et al. (2020) underscore the importance of understanding the cyberattack life cycle and mapping out potential threats using AI-based frameworks. By categorizing malicious uses of AI and identifying key limitations in existing cybersecurity landscapes, researchers aim to develop proactive defense mechanisms that can anticipate and mitigate emerging threats.

Furthermore, the integration of big data analytics and machine learning holds promise for enhancing the efficiency and effectiveness of cybersecurity operations. Nassar and Kamal highlight the significance of machine learning in enabling adaptive security postures that can dynamically respond to evolving threats. Similarly, Chehri et al. (2021) propose qualitative risk assessment methods that leverage big data and AI to identify vulnerabilities and recommend suitable security countermeasures for critical infrastructures like smart grids.

The discussions by Tornjanski et al. (2021), Rana and Rana (2021), Bai and Fang (2022), and An (2022) emphasize the importance of organizational culture and collaboration in building resilient cybersecurity ecosystems. Proactive security measures, continual learning, and human oversight are identified as critical components of a comprehensive cybersecurity strategy. Moreover, ethical considerations surrounding the use of AI in cybersecurity are also highlighted, with an emphasis on transparency, bias mitigation, and responsible governance.

6.10 FINDINGS

The discussion presented above reveals a number of significant findings.

The dynamic nature of cyber threats: Cyber threats continue to evolve in complexity and sophistication, necessitating proactive and adaptive security measures. The role of AI and big data analytics: AI and big data analytics offer promising avenues for enhancing cybersecurity through real-time threat detection, anomaly identification, and adaptive responses.

Framework development: Researchers have developed frameworks for categorizing AI-based cyberattacks, mapping out the cyberattack life cycle, and identifying key limitations in existing cybersecurity landscapes.

Organizational culture and collaboration: Proactive collaboration between stakeholders, continual learning, and sound data governance are essential for building resilient cybersecurity ecosystems.

Ethical considerations: Ethical considerations surrounding the use of AI in cybersecurity, including issues of privacy, bias mitigation, and transparency, are of paramount importance.

6.11 SUGGESTIONS

Based on the findings, several suggestions can be made for future research and practice in cybersecurity:

Further research on AI-based cyberattacks: Continued research is needed to explore emerging threats and develop effective countermeasures against AI-based cyberattacks.

Enhanced collaboration: Stakeholders should collaborate proactively to share threat intelligence, best practices, and resources for combating cyber threats.

Investment in AI and big data analytics: Organizations should invest in AI and big data analytics capabilities to bolster their cybersecurity defenses and stay ahead of evolving threats.

Ethical guidelines and governance: Clear ethical guidelines and governance frameworks should be established to ensure the responsible and transparent use of AI in cybersecurity operations.

6.12 LIMITATIONS

Despite the valuable insights provided by the reviewed studies, there are several limitations to consider:

Generalizability: The findings may not be universally applicable across different organizational contexts and industries.

Emerging technologies: The rapid pace of technological innovation may render some findings outdated or incomplete.

Data limitations: The availability and quality of data may pose challenges for conducting robust analyses and developing effective cybersecurity solutions.

Ethical considerations: Ethical considerations surrounding the use of AI in cybersecurity operations are complex and multifaceted, requiring careful consideration and ongoing debate.

6.13 CONCLUSION

In conclusion, the integration of AI, big data analytics, and machine learning holds tremendous promise for enhancing cybersecurity in the face of evolving threats. By understanding the cyberattack life cycle, developing proactive defense mechanisms, and fostering collaboration among stakeholders, organizations can build resilient cybersecurity ecosystems capable of withstanding emerging threats. However, ethical considerations surrounding the use of AI in cybersecurity operations must be carefully addressed to ensure responsible and transparent governance. Moving forward, continued investment in research, collaboration, and technological innovation will be essential for staying ahead of the curve in the ongoing battle against cybercrime.

REFERENCES

An, A. (2022). The evolution of cyber security threats in the digital age. *International Journal of Business Management and Visuals*, 5(2), 22–29.

Azambuja, A. J. G., Plesker, C., Schützer, K., Anderl, R., Schleich, B., & Almeida, V. R. (2023). Artificial intelligence-based cyber security in the context of industry 4.0—A survey. *Electronics*, 12(8), 1920.

Bai, M., & Fang, X. (2022). Cybersecurity analytics: AI's role in big data threat detection. *Eduzone: International Peer Reviewed/Refereed Multidisciplinary Journal*, 11(2), 392–396.

Chehri, A., Fofana, I., & Yang, X. (2021). Security risk modeling in smart grid critical infrastructures in the era of big data and artificial intelligence. *Sustainability*, 13(6), 3196.

George, A. S. (2023). Securing the future of finance: How AI, Blockchain, and machine learning safeguard emerging Neobank technology against evolving cyber threats. *Partners Universal Innovative Research Publication*, 1(1), 54–66.

Kaloudi, N., & Li, J. (2020). The AI-based cyber threat landscape: A survey. *ACM Computing Surveys (CSUR)*, 53(1), 1–34.

Kumar, S., Gupta, U., Singh, A. K., & Singh, A. K. (2023). Artificial intelligence: revolutionizing cyber security in the digital era. *Journal of Computers, Mechanical and Management*, 2(3), 31–42.

Rana, S. K., & Rana, S. K. (2021). Intelligent amalgamation of blockchain technology with industry 4.0 to improve security. In *Internet of Things* Arun Kumar Rana, Ayodeji Olalekan Salau, Sharad Sharma, Shubham Tayal, Swati Gupta (eds.) (pp. 165–175). CRC Press.

Rangaraju, S. (2023). AI sentry: Reinventing cybersecurity through intelligent threat detection. *EPH-International Journal of Science And Engineering*, 9(3), 30–35.

Rawat, D. B., Doku, R., & Garuba, M. (2019). Cybersecurity in big data era: From securing big data to data-driven security. *IEEE Transactions on Services Computing*, 14(6), 2055–2072.

Samtani, S., Kantarcioglu, M., & Chen, H. (2020). Trailblazing the artificial intelligence for cybersecurity discipline: a multi-disciplinary research roadmap. *ACM Transactions on Management Information Systems (TMIS)*, 11(4), 1–19.

Tornjanski, V., Knežević, S., Ljubanić, D., Glišić, V., Žižić, D., & Travica, J. (2021, September). Towards secured digital business ecosystems: From threats to opportunities. In *E-business technologies conference proceedings* (Vol. 1, No. 1, pp. 1–14).

Chapter 7

Integration of blockchain technology with travel industry

A comprehensive analysis

Sumit Kumar Rana, Navneet Verma and Arun Rana

7.1 INTRODUCTION

Although blockchain technology may appear complicated at first, it is actually extremely simple to grasp. It is a collection of public records, often known as a public ledger that lists or records transactions between parties [1–4]. Cryptography protects each record, also known as a 'block' in blockchain language. One of the most fundamental aspects of blockchain technology is the distribution of data over a peer-to-peer network. Figure 7.1 illustrates that each block contains transaction details as well as a timestamp. Blocks are also permanent and cannot be modified without gaining consensus from the whole network and affecting all subsequent blocks [5]. Because it is decentralized, permanent, time-stamped, and immutable, blockchain data is more secure, traceable, and transparent. There is no single point of vulnerability or failure and the data is resistant to alteration and unauthorized access. According to Fortune Business Insights' Blockchain Technology Industry Report, the global blockchain technology industry would grow at a CAGR of 59.9% through 2030.

The travel industry relies on multiple businesses providing information. For example, travel agents must send customer information to hotels and airlines, and consumers' personal belongings are regularly transferred and tracked across firms. Because the responsibility for data preservation is distributed across the whole network, data access and storage becomes very reliable and easy [6].

Of course, the travel industry relies significantly on financial transactions, and blockchain technology has the potential to simplify and secure payments. This is particularly true when dealing with cash from outside the country. As a result, blockchain has the ability to boost confidence in all stakeholders [7].

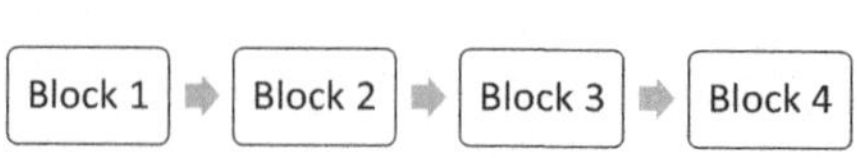

Figure 7.1 Blockchain technology.

DOI: 10.1201/9781003590439-7

Blockchain technology has revolutionized the travel and hospitality industries. It has modified the booking and transaction methods. Blockchain professionals in various firms leverage the benefit of Blockchain development to reinforce their functioning mechanism in travel and tourism [8].

This research study aims to provide an in-depth analysis of the potential benefits, challenges, current applications, and prospects of blockchain adoption in the tourism industry. The article seeks to elucidate the disruptive impact of blockchain technology on various aspects of the travel business, including data security, payment systems, supply chain management, and customer experience [9–11]. By reviewing existing literature, analyzing case studies, and examining current trends, the paper intends to offer valuable insights for industry professionals, researchers, and policymakers interested in understanding blockchain's role in reshaping business models, enhancing operational efficiency, and building trust within the travel ecosystem. The research will explore the fundamental principles of blockchain technology, identify key challenges and limitations associated with its implementation in the travel industry, examine current applications and initiatives, and predict future trends and opportunities for blockchain adoption. Overall, the research paper aims to advance our understanding of how blockchain technology could transform the travel sector, leading to a safer, more transparent, and efficient travel ecosystem.

7.2 LITERATURE SURVEY

This [11] study used theoretical, methodological, and systematic source analysis approaches, such as statistical materials, regulatory documents, and scientific articles, to examine the current state of blockchain technology use in tourism and its potential applications. The following findings were obtained: the main directions of blockchain application in tourism were recognized, as well as its impact on the development of a more diverse and flexible offer in the tourism services market. It is decided that blockchain technology can give improved security and transparency in the tourism business, enhancing revenue, which in turn influences job generation and the population's level of living.

This [12] study presented few research suggestions and recommendations to help with the development of blockchain-based tourism experiences, while also taking into account social barriers through a detailed contextualization of the travel industry. Indeed, it is possible that Blockchain will need to adapt to the tourism industry and its associated sustainability concerns, rather than the other way around.

This [13] research paper, which focuses on sustainability and responsible tourism, looks into the use of blockchain technology in supporting ecologically friendly practices in the travel industry. The article investigates blockchain-based solutions for carbon offset programs, eco-certifications, and transparent supply chains, and analyzes their potential to encourage sustainable behavior among travelers and service providers. It also investigates regulatory and technological barriers to the deployment of blockchain for sustainable tourism initiatives.

The goal of this [14] study was threefold: first, to identify the barriers to blockchain adoption using a literature review and expert viewpoints. The proposed rough Interpretive Structural Modelling - Cross-Impact Matrix Multiplication, based on expert opinions, should then be utilized to evaluate them. Finally, consider how these issues connect to the diffusion of innovations theory. The findings suggested that "lack of technical maturity" and "lack of interoperability" were the most major blockchain challenges in the tourism industry. The outcomes of the study supported macro- and micro-level decision-making in the tourism industry's prospective blockchain applications.

Technological breakthroughs have historically facilitated the development of new business models in the tourism industry [15], incorporating modernization processes while also introducing new technological developments associated with the marketing and distribution of tourism services. Technological developments have played and will continue to play an important role in the tourism industry, which is increasingly concerned about sustainability. This worry is mostly related to public policies aiming at assessing tourism demand, which have implications for parties involved in tourism activities. It also has an environmental impact that jeopardizes the balance of natural ecosystems [16].

This [17] article suggested that transparency provided by the distributed ledger can improve the trust level of end users in the travel ecosystem. As this emerging technology known as Blockchain technology works in decentralized manner so middlemen can be easily removed from the system. Travel industry must understand the advantages that distributed ledger technology may offer, such as greater client loyalty and a competitive advantage. There are presently various blockchain-based online travel platforms available, and this is simply the foundation of the change. Today, this technology remains extremely technical, making it inaccessible to tourist specialists.

Abrupt technological changes are changing the business models of different industries. Same transformation is visible in travel industry also. Now everything is available online but current system works in centralized manner. Blockchain technology can help in building a decentralized ecosystem for the travel industry. End user will be the owner of his personal data and nobody will be able to access without his permission [18].

Technology is used extensively in all aspects of tourism, whether it is by the traveler or the travel service provider. The former can use it to locate, personalize, and purchase travel-related items, while the latter can use it to create, oversee, and market their services internationally [19].

Due to the intricacy of the tourism industry, the blockchain concept can be implemented at multiple pivotal points throughout it. The implementation of blockchain technology in tourism establishments will facilitate the management of transactions pertaining to sales, operations, finance, and administration, thereby transforming the centralized tourism ecosystem [20]. Since blockchain technology is still in its infancy in the travel and tourism sector, it can be used as an educational tool through a range of applications that can be developed.

Blockchain technology can serve as the foundation for applications that aid in the determination of a person's identity, which is a significant advantage for those who travel frequently and must produce identification documents [21]. Customer reviews are critical in the tourism sector, but they must be genuine for a private key to be issued and signed using blockchain technology. In recent years, secrecy and data protection have been fiercely disputed issues, with efforts being made to discover immediate and long-term solutions. Blockchain technology has numerous uses for managing data in an integrated manner while protecting a person's identity.

Study of existing work provides various insights for better understanding of the domain. Various researchers have discussed the application areas that can be improved by utilizing the features of the blockchain technology. Some of the potential benefits of incorporating blockchain technology in travel industry are discussed in the next section.

7.3 POTENTIAL BENEFITS OF BLOCKCHAIN IN THE TRAVEL INDUSTRY

There are multiple benefits that can be achieved by integrating blockchain technology holds significant promise for revolutionizing the travel industry by offering a multitude of potential benefits:

1. Data Security: blockchain technology uses a distributed ledgers to record all the transactions. Once the transaction has been written on the ledgers, nobody can delete or update that particular transaction. So, this immutability feature of the ledger maintains the security of the information.
2. Peer to peer network: blockchain uses a peer-to-peer network for communication among the nodes of the network. It does not require any third party to communicate. Removal of intermediaries makes the communication faster and cost effective. It uses the smart contracts, which is a programmed contract, for the settlement of the payments and minimizes the chances of disputes.
3. Improved Supply Chain Management: by using distributed and immutable ledger, all the events of journey like ticket booking, luggage tracking, accommodation, etc., can be stored. This brings transparency, trust and efficiency in the travel ecosystem.
4. Enhanced Customer Experience and Loyalty: Blockchain technology can be utilized for creating decentralized identity for customers. These

blockchain based identities can help the customers for various activities like simplifying check-in processes, security screenings, and loyalty program enrolments.

5. Prevention of Fraudulent Activities: distributed ledger in blockchain in immutable. Once the transaction is written on it, it cannot be deleted or modified. This attribute prevents the fraudulent activities such as ticket scalping, identity theft, or counterfeit bookings.
6. Environmental Sustainability: Blockchain technology will automate the complete process and convert it into a paperless process. It will lead to reduced carbon emissions and minimize the environmental effects of conventional process.

These abovementioned benefits make it a suitable choice for the travel industry. But every coin has two sides. There are few challenges and limitations with the adoption of this emerging technology which are discussed in the next section.

7.4 CHALLENGES AND LIMITATIONS

While the integration of blockchain technology in the travel industry offers numerous potential benefits, it also presents several challenges and limitations that need to be addressed as shown in Figure 7.2:

1. Scalability Issues: in case of travel industry, millions of transactions get executed daily across the world. These millions of transactions cannot be handled by public blockchains like bitcoin, Ethereum. Speed of these blockchains are approximately 10–20 transactions per second. This can raise a scalability issue that can impact the adoption of blockchain technology.
2. Regulatory and Legal Frameworks: blockchain technology is an emerging technology so the standardization and legal frameworks are under development and in evolving stage. So, various stakeholders and policy makers mutual agreement about the regulatory problems is in confusing state.
3. Interoperability: interoperability of the conventional system with the decentralized system is complex and in inception state. Incorporation of this emerging technology with the current systems requires a lot of planning and investment.
4. Environmental Impact: mining process of blockchain consume a lot of energy, particularly in proof-of-work consensus mechanisms like Bitcoin. It can neutralize its benefits of paperless process. This energy consumption requires a careful consideration before integrating it with the traditional system.
5. Perception and Trust Issues: Adoption of any new technology faces resistance initially. So to break the perception and build the trust among the stakeholders for the technology is a challenge.

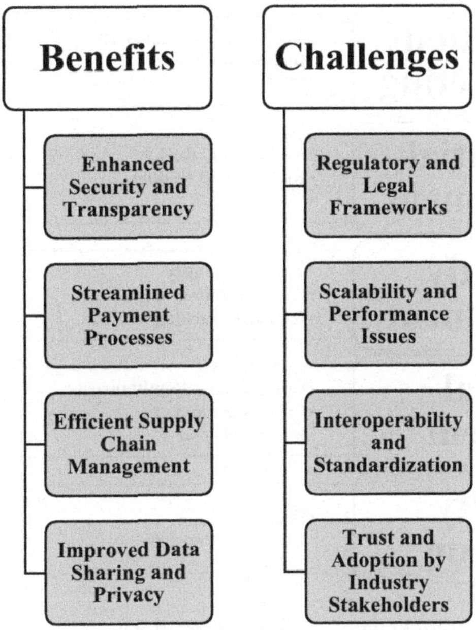

Figure 7.2 Benefits and challenges of blockchain in tourism.

6. Cost and Resource Requirements: whenever any new technology replaces any conventional system, it requires investment for the transformation. Finding skilled human resources for an emerging technology is also challenging. Small travel businesses might not be in favor of this transformation because of the cost involved in it.

Integration of any emerging technology with the traditional systems faces different challenges and resistance from the market. These challenges can be handled collectively by the collaboration of all the stakeholders. Next section will discuss the applications of blockchain technology in the travel industry.

7.5 CURRENT APPLICATIONS OF BLOCKCHAIN IN TRAVEL

There are multiple spaces in the travel industry where blockchain technology can be applied to improve the process as shown in Figure 7.3. Some of the applications are discussed below:

1. Smart Contracts for Ticketing and Bookings: blockchain technology supports self-executing codes known as smart contracts. These self-executing codes can automate various processes without human intervention. This can help in eliminating the middlemen and prevent the chances of fraudulent activities.

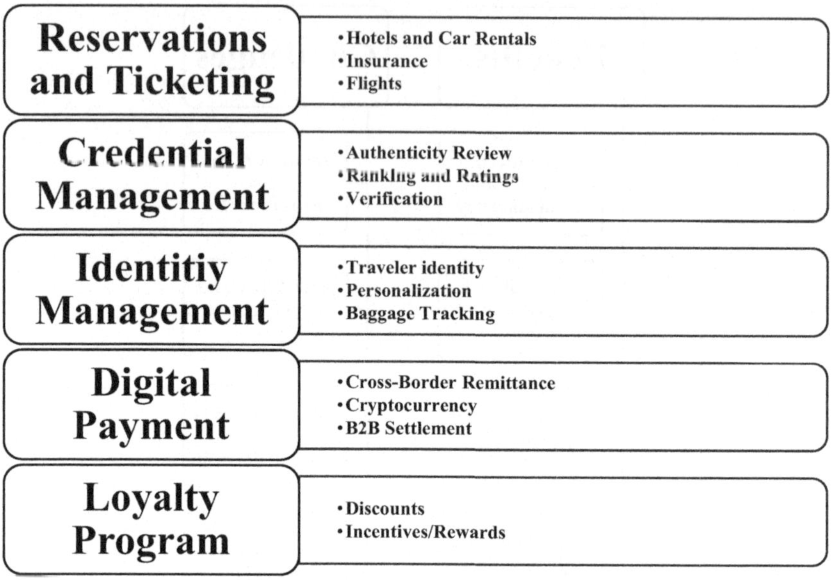

Figure 7.3 Usage of blockchain in tourism industry.

2. Decentralized identity: Blockchain can help in creating decentralized identities for all the customers. These decentralized identities will help the customer to safeguard their data. In centralized system, customers personal data be misused for various fraudulent activities. Decentralized identities can be used at various steps like check in and check out at airport, hotels, etc., and at the same time nobody will be able to use the personal data of the customer without his permission.
3. Loyalty Programs and Rewards: different travel businesses can run loyalty programs using blockchain technology. These programs will be transparent, flexible and secured from the perspective of the customer. Distributed ledger helps in avoiding the fraud and manipulation with the data.
4. Supply Chain Transparency: distributed ledger which is immutable in nature help in enhancing the transparency and traceability in the travel supply chain. This ledger records every transaction and these transactions cannot be manipulated for illegitimate benefits. Every product or service can be traced and tracked for auditing purposes.
5. Peer-to-Peer Travel Services: blockchain technology supports peer to peer communication so the service providers and service requester can communicate directly without the involvement of any intermediaries. Removal of middlemen can help in saving the time and money of the customer. No single entity will be able to control the network because network is not owned by any single entity but controlled by all the participants of the network.

7.6 RECOMMENDATIONS FOR STAKEHOLDERS

There are various ways that stakeholders can opt to promote blockchain and travel industry integration. Some of these ways are discussed below:

1. Education and Awareness campaign: blockchain technology is an emerging technology so is less known among the community. Different awareness campaign should be run to spread awareness among different stakeholders about the features and benefits of the technology. Different workshops and seminars can be organized about adoption of emerging technologies.
2. Collaborate and Form Partnerships: there are multiple stakeholders like travel businesses, regulatory bodies, customers, etc., these stakeholders should collaborate for resource sharing and knowledge exchange so that a better ecosystem can be built for the integration of blockchain technology and conventional travel industry system.
3. Conduct Pilot Projects: as the transformation from the traditional system to new decentralized system is not a game of few days. This is long time transformation that will take place in different phases. Before a complete switch over of the system, pilot projects should be run and tested for the feasibility and effectiveness of the new solution. Then stakeholders can evaluate the benefits and challenges of the new systems.
4. Ensure Regulatory Compliance: emerging technology come with its pros and cons. Before the integration of the technology, all the stakeholders must consult with their legal advisors and regulatory bodies to understand the complexities in the regulations and compliance of the system.
5. User Experience and Adoption: to adopt, any new technology one of the most important things is user experience. Proper testing should be conducted to ensure that interface is user friendly and intuitive. Feedbacks from stakeholders should be collected for proper evaluation before adoption of the new technology.

7.7 CONCLUSION

Tourism industry is one of the biggest global industry which is present all around the world. This industry includes various stakeholders like hotels, restaurants, transportation, and other service providers. When there are multiple businesses are involved in any task then main concern is about the trust and transparency. Traditional ecosystem suffers with issues like manipulation of data, data privacy, trust, presence of intermediaries. These issues can be solved by integrating blockchain technology with the conventional ecosystem. Blockchain technology will help in building a decentralized ecosystem where no single entity will be able to control the ecosystem. All the activities will be transparent in this decentralized ecosystem which will raise the trust level of stakeholders. All the transactions will be immutable so no entity will be able to manipulate the data for his personal benefits.

There are some challenges and limitations with this new decentralized ecosystem. First of all, the adoption of any new technology always faces the resistance. Nobody wants to replace their legacy system with the new technology initially. There are no regulatory body that controls and develops the regulatory framework. Without the presences of regulatory body, stakeholders will hesitate in integration of this new technology. Another issue will be of scalability. Currently available blockchain technology can process 10–20 transactions per second but in travel industry millions of transactions are executed simultaneously. Despite these challenges, researchers are continuously exploring and investing for the integration of blockchain technology with the traditional travel ecosystem. As the development will progress, many of the challenges and limitations will be eliminated and integration of the blockchain with the traditional business models will grow.

REFERENCES

[1] Melkić, Sara, and Nevenka Čavlek. "The impact of blockchain technology on tourism intermediation." *Tourism: An International Interdisciplinary Journal* 68, no. 2 (2020): 130–143.

[2] Rana, Roberto Leonardo, Nino Adamashvili, and Caterina Tricase. "The impact of blockchain technology adoption on tourism industry: A systematic literature review." *Sustainability* 14, no. 12 (2022): 7383.

[3] Subramanian Nachiappan, Chaudhuri Atanu, and Kayıkcı Yaşanur. "Blockchain applications and future opportunities in transportation." *Blockchain and Supply Chain Logistics: Evolutionary Case Studies* (2020): 39–48.

[4] Khanna, Abhirup, Anushree Sah, Tanupriya Choudhury, and Piyush Maheshwari. "Blockchain technology for hospitality industry." In *Information Systems: 17th European, Mediterranean, and Middle Eastern Conference, EMCIS 2020, Dubai, United Arab Emirates, November 25–26, 2020, Proceedings 17*, pp. 99–112. Springer International Publishing, 2020.

[5] Sharma, Mahak, Rajat Sehrawat, Tugrul Daim, and Amir Shaygan. "Technology assessment: Enabling blockchain in hospitality and tourism sectors." *Technological Forecasting and Social Change* 169 (2021): 120810.

[6] Rana, S.K., Rana, A.K. and Dhawan, S. A vital fusion of Internet of medical things and blockchain to transform data privacy and security. In *Convergence of Deep Learning and Artificial Intelligence in Internet of Things* Ajay Rana, Arun Kumar Rana, Sachin Dhawan, Sharad Sharma, Ahmed A Elngar (eds.) (pp. 293–308). CRC Press, 2022.

[7] Ozdemir, Ali Ihsan, Ilker Murat Ar, and Ismail Erol. "Assessment of blockchain applications in travel and tourism industry." *Quality & Quantity* 54 (2020): 1549–1563.

[8] Erceg, Aleksandar, Jovanka Damoska Sekuloska, and Ivan Kelić. "Blockchain in the tourism industry—A review of the situation in Croatia and Macedonia." *Informatics* (vol. 7, no. 1, p. 5). MDPI, 2020.

[9] Calvaresi, Davide, Maxine Leis, Alevtina Dubovitskaya, Roland Schegg, and Michael Schumacher. "Trust in tourism via blockchain technology: Results from a systematic review." In *Information and Communication Technologies in Tourism 2019: Proceedings of the International Conference in Nicosia, Cyprus, January 30–February 1, 2019*, pp. 304–317. Springer International Publishing, 2019.

[10] Rana, S.K., S.K. Rana, A.K. Rana, and S.M. Islam A blockchain supported model for secure exchange of land ownership: an innovative approach. In *2022 International Conference on Computing, Communication, and Intelligent Systems (ICCCIS)* (pp. 484–489). IEEE, 2022, November.

[11] Rashideh, Waleed. "Blockchain technology framework: Current and future perspectives for the tourism industry." *Tourism Management* 80 (2020): 104125.

[12] Haponenko, Hanna, and Vladyslav Vasylenko. "Prospects for the use of blockchain technology in the tourism industry." *The Journal of VN Karazin Kharkiv National University. Series: International Relations. Economics. Country Studies. Tourism* 10 (2019): 193–199.

[13] Fragnière, Emmanuel, Jean-Michel Sahut, Lubica Hikkerova, Roland Schegg, Michael Schumacher, Sandra Grèzes, and Randolf Ramseyer. "Blockchain technology in the tourism industry: New perspectives in Switzerland." *Journal of Innovation Economics Management* 37, no. 1 (2022): 65–90.

[14] Rana, S.K., A.K. Rana, S.M.I. Naz, A.K. Singh, and B. Bhushan Supply chain strategies supported by blockchain technology: open problems and challenges. In *2023 International Conference on Advancement in Computation & Computer Technologies (InCACCT)* (pp. 1–6). IEEE, 2023, May.

[15] Erol, Ismail, Irem Onder Neuhofer, Tarik Dogru, Ahmet Oztel, Cory Searcy, and Ali C. Yorulmaz "Improving sustainability in the tourism industry through blockchain technology: Challenges and opportunities." *Tourism Management* 93 (2022): 104628.

[16] Rana, S. K., & Rana, S. K. (2021). Intelligent amalgamation of blockchain technology with industry 4.0 to improve security. In *Internet of Things* Arun Kumar Rana, Ayodeji Olalekan Salau, Sharad Sharma, Shubham Tayal, Swati Gupta (eds.) (pp. 165–175). CRC Press.

[17] Erceg, Aleksandar, Jovanka Damoska Sekuloska, and Ivan Kelić "Blockchain in the tourism industry—A review of the situation in Croatia and Macedonia." *Informatics*, (vol. 7, no. 1, p. 5). MDPI, 2020.

[18] Colombo, Edoardo, and Rodolfo Baggio. "Tourism distribution channels." In Knowledge transfer to and within tourism (vol. 8, pp. 289–301). Emerald Publishing Limited, 2017.

[19] Rana, S. K., S. K. Rana, A. K. Rana, K. Nisar, T. R. Soomro, & S. Nisar A survey on blockchain technology supported approaches for healthcare system, open issues and challenges. In *2022 14th International Conference on Mathematics, Actuarial Science, Computer Science and Statistics (MACS)* (pp. 1–7). IEEE, 2022, November.

[20] Treiblmaier, Horst. "Blockchain and tourism." In Handbook of e-Tourism, pp. 475–495. Cham: Springer International Publishing, 2022.

[21] Fragnière, Emmanuel, Jean-Michel Sahut, Lubica Hikkerova, Roland Schegg, Michael Schumacher, Sandra Grèzes, and Randolf Ramseyer. "Blockchain technology in the tourism industry: New perspectives in Switzerland." *Journal of Innovation Economics Management* 37, no. 1 (2022): 65–90.

Chapter 8

Intelligent data mining techniques

RBF neural network-based intelligent optimal control and adaptive bound for collaborative robot manipulators grasping an object

Komal Rani, Naveen Kumar, Manju Rani and Kamal

8.1 INTRODUCTION

Research on coordinated control of several robot manipulators has received increased interest.

In many industrial applications, it is of great significance due to versatility in tasks done by multiple robots i.e. transporting materials, handling of large and heavy objects, dismantling bombs etc. Such type of tasks may not be achieved successfully by single robot manipulators. However, the controllers design for cooperative multiple robot manipulators is intricate than that of single manipulators. The primary complexity arises from the fact that, when carrying an object, several robot manipulators combine to create a closed chain mechanism that is extremely nonlinear and linked. Furthermore, the presence of internal pressures that are essential for preserving the correct contact between the manipulator end effector and the item but do not impact the object's motion makes this problem more complex.

The hybrid force/position controller and the impedance control approach form the basis of the majority of the control scheme used to solve this issue. [1, 2] presents a master slave control mechanism for many manipulators. The hybrid force/position control method was created by Hayati [3] to coordinate multi-arm control. In order to create a robust controller, Tarn et al. [4] explored the theory of linear optimum control. Uchiyama and Dauchez [5] used the set of static and kinematic equations generated for dual manipulators to construct a symmetric hybrid force/position control system.

Other model based control methods can be noted in [6–8]. But complete knowledge of complex dynamics is required for the success of aforementioned control problem. The system's uncertainty causes the controller's presentation to decrease and the system to become unstable.

Then to overcome the problems occur due to model based control scheme, several adaptive methods with updation of robot parameters have been seen in the literature. Hu et al. [9] offered a flexible method to offset robotic system uncertainties based on Popov hyperstability theory.

DOI: 10.1201/9781003590439-8

Walker and Kim [10] addressed methodical adaptive control for two robot manipulators carrying an unspecified payload. To regulate the position and force of the robot manipulators grabbing an object, Yao et al. [11] created a variable structure controller. A flexible approach to managing several manipulators gripping a single object was introduced by Jean [12]. This control scheme is developed without the contact forces. Gueaieba and Sharhanb [13] developed robust decentralized method to control the internal forces and position simultaneously for cooperative multiple robot manipulators with uncertainties. Other adaptive methods can be noted in [14–16]. But, these adaptive method requires the computation of regression matrix separately using the system model which is not an easy task.

Recently, in order to develop the model free controllers, more work on neural network controllers can be noticed in the literature [17–21] due to ability of NN to approximate the uncertainties through self-learning in order to solve the control problem of coordinating two redundant robots, Hou & Cheng [22] employed a dual neural network approach. Siqueira [23] used H infinity control scheme based on NN technique for fully actuated and under-actuated cooperative robot manipulators. A neural network-based adaptive method employing optimal back propagation was created by Panwar et al. [24] to enable coordinated motion of many manipulators without the effects of friction or outside disturbances. The neural network based force/position control method was evolved for coordinate non holonomic mobile manipulator [25]. For coordinated robotic manipulators, Zhang et al. [26] suggested adaptive NN control as a way to increase system stability and reduce vibration.

But above mentioned control schemes do not optimize the long term performance function and are not optimal. So, for optimization, among various intelligent control methods, optimal neural network control scheme have been addressed by [26–32] for robotic system like fixed base manipulators, planer manipulators and discrete-time system. But in case of multiple manipulators, this approach is found very limited in literature review. Sadati and Babazadeh [33] presented an optimum control of multiple-arm robots by decomposition coordination framework of large-scale systems. Rani [34] designed an optimal control scheme with neural network and adaptive bound system for multiple mobile manipulators. A reinforcement learning technique for the best coordination control strategy for several robots handling a shared object and tracking the trajectory was discussed by Li et al. [35]. Sieber et al. [36] developed LQR optimal control for cooperative manipulators under some constraints to maintain its formation and to regulate configuration of the multi-robot team but dependency on iterative scheme reduces the quality of the controller.

The fact that system error converges to a limited neighborhood of zero is a crucial remark about the optimum control approaches mentioned above. Due to the highly nonlinear and coupled mechanism of several robot manipulators operating in the presence of uncertainties and while gripping a shared object. This encourages the author to create an intelligent, asymptomatically stable optimum control strategy for the position and force control of many manipulators working together to grab a common item. The combination of RBFNN, adaptive bound term, and linear optimum control scheme results in the generation of intelligent optimal controller. There is a spectrum of system uncertainties that the linear optimum controller is resilient to. Nevertheless, the nonlinear dynamics of the system are not taken into consideration when constructing the linear optimal-controller. We estimated the system's nonlinearity using RBF neural networks, and we estimated the uncertainties and unknown bounds on the NN reconstruction error using the adaptive bound component.

Furthermore, the proposed controller has the following salient features: (i) A priori knowledge of system dynamics is used for designing linear optimal-control; (ii) The HJB equation is no longer required for creating the control scheme since the Riccati equation must be solved explicitly for the HJB equation to yield optimal control; (iii) RBFNN is employed to estimate nonlinearity of the system and adaptive robust part of proposed controller approximates reconstruction errors of neural network and unknown bounds on unstructured uncertainties; (iv) In contrast to conventional optimum control strategies, which converge error to a limited neighborhood around zero, the system asymptotically converges to zero.

8.2 MODELING OF THE SYSTEM

8.2.1 System description

Imagine the system depicted in Figure 8.1—a coordinated system of several robot manipulators handling a shared object at designated points. For systematic modeling, we define the different coordinate frame as follows: $OX_W Y_W Z_W$ = World coordinate frame. $OX_O Y_O Z_O$ = Object coordinate frame. $OX_{Bi} Y_{Bi} Z_{Bi}$ = Base coordinate frame for ith robot manipulator. $OX_{Ei} Y_{Ei} Z_{Ei}$ = End effector coordinate frame for ith robot manipulator. To frame the system's dynamics, it is presumable that (i) the object is rigid, (ii) manipulator firmly grasps the object, and (iii) the manipulator's end effector and the object are not moving relative to one another.

8.3 DYNAMICS OF COORDINATED MULTIPLE ROBOT MANIPULATORS

Lagrangian mechanics-based general equation can be used to show the dynamics of the ith robot manipulator. [24]:

$$D_i(q_i)\ddot{q}_i + B_i(q_i,\dot{q}_i)\dot{q}_i + G_i(q_i) + F_i(\dot{q}_i) + T_{di} = \tau_i - J_i^T f_i,\ i = 1,2,\ldots k. \quad (8.1)$$

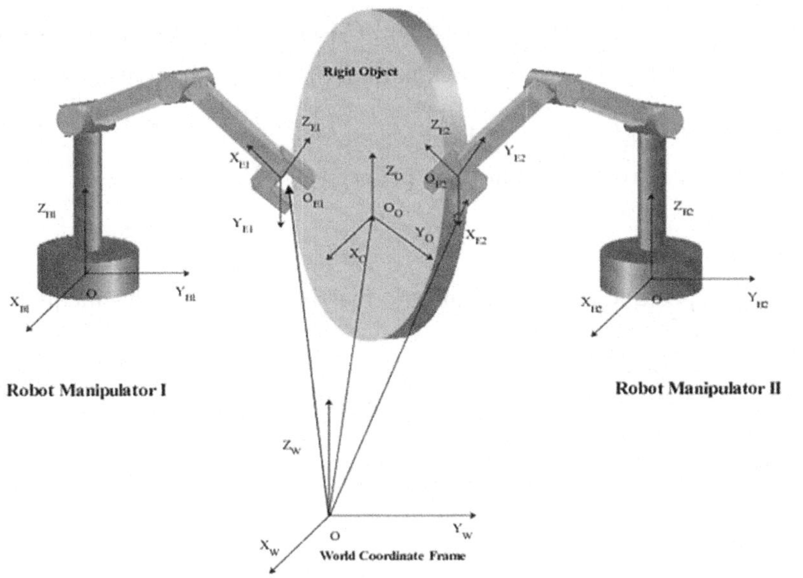

Figure 8.1 Structure of coordinated multiple robot manipulators grasping an object.

where $q_i \in \mathbb{R}^n$ is the joint variable, $D_i(q_i) \in \mathbb{R}^{nxn}$ stands for the inertia matrix, $B_i(q_i,\dot{q}_i) \in \mathbb{R}^{nxn}$ represents the centripetal-coriolis matrix, $G_i(q_i) \in \mathbb{R}^n$ is the gravity term, $T_{di} \in \mathbb{R}^n$ is the disturbance term, $F_i(\dot{q}_i) \in \mathbb{R}^n$ represents friction effects, $\tau_i \in \mathbb{R}^n$ is the torque input, $J_i \in \mathbb{R}^{m\times n}$ is a full rank jacobian matrix and $f_i = \left[\kappa_i^T, \mu_i^T\right] \in \mathbb{R}^n$ is the vector of force or moment applied to the item by the ith robot manipulator. The following criteria are satisfied by the ith robot dynamics (1):

Property 1

Positive definite symmetric matrix D_i (q_i) satisfies the following inequality:

$$\alpha_{Di} I_n \le D_i(q_i) \le \beta_{Di} I_n \tag{8.2}$$

where α_{Di} and β_{Di} are positive constants.

Property 2

The relation $\left(\dot{D}_i(q_i) - 2B_i(q_i,\dot{q}_i)\right)$ is skew symmetric.

Given by (8.1), the dynamics of k robot manipulators can be compactly represented as:

$$D(q)\ddot{q} + B(q,\dot{q})\dot{q} + G(q) + F(\dot{q}) + T_d = \tau - J^T(q)f, \tag{8.3}$$

where $q = \left[q_1^T,\cdots,q_k^T\right]^T$, $D(q)$ =blockdiag$\left[D_1(q),\cdots,D_k(q)\right]$, $B(q,\dot{q})$ =blockdiag $\left[B_1(q,\dot{q}),\cdots,B_k(q,\dot{q})\right]$, $J(q)$ = blockdiag $\left[J_1(q),\cdots,J_k(q)\right]$, $G(q)=\left[G_1^T(q),\cdots,G_k^T(q)\right]^T$, $F(\dot{q})=\left[F_1^T(\dot{q}),\cdots,F_k^T(\dot{q})\right]^T$, $T_d=\left[T_{d1}^T,\cdots,T_{dk}^T\right]^T$, $\tau=\left[\tau_1^T,\cdots,\tau_k^T\right]^T$. Also, Equation (3) satisfies the following properties [24]:

Property 3

$D(q)$ is positive definite symmetric matrix, which satisfies inequality given as:

$$\alpha_D I_n \le D(q) \le \beta_D I_n \tag{8.4}$$

where α_D and β_D are positive ceonstants.

Property 4

The relation $\left(\dot{D}(q)-2B(q,\dot{q})\right)$ of matrices is skew symmetric.

8.3.1 Force and common object dynamics

Each robot manipulator exerts force/moment $f_i=\left[\kappa_i^T,\mu_i^T\right]$ on the objects. Then the resultant moment\force $\boldsymbol{K_o}=\left[\kappa_o^T,\mu_o^T\right]$ on the grasped objects is given by

$$\kappa_o=\sum_{i=1}^{k}\kappa_i \tag{8.5}$$

$$\mu_o=\sum_{i=1}^{k}\left(\mu_i+p_{OE_i}*\kappa_i\right) \tag{8.6}$$

Equations (8.5) and (8.6) can be written in a single form as

$$K_o=\sum_{i=1}^{k}\Omega_i f_i \tag{8.7}$$

where $\Omega_i=\begin{bmatrix} I & 0 \\ T_{p\kappa} & I \end{bmatrix}$ is a matrix grab form at the ith point of contact with $T_{p\kappa}=p_{OE_i}*\kappa_i$ and I as identity matrix. Equation (8.7)'s compact form can now be expressed as

$$K_o=\Omega f \tag{8.8}$$

where $\Omega = \left[\Omega_1,\cdots,\Omega_k\right]$ and $f = \left[f_1^T,\cdots,f_k^T\right]^T$. Moreover, Equation (8.8) may be written as [24]:

$$f = \Omega^+ K_o + \left(I - \Omega^+ \Omega\right)\eta \tag{8.9}$$

where $\Omega^+ = \Omega^T\left(\Omega\Omega^T\right)^{-1}$ is the pseudo inverse of Ω and η represents arbitrary vector of nk dimension which is referred as auxiliary internal force κ_{int}.

The dynamics of the common rigid object grasped by multirobot manipulators can be expressed as [24]

$$D_o\left(v_o\right)\ddot{v}_o + B_o\left(v_o,\dot{v}_o\right)\dot{v}_o + G_o\left(v_o\right) = K_o, \tag{8.10}$$

where $D_o\left(v_o\right), B_o\left(v_o,\dot{v}_o\right)$ and $G_o\left(v_o\right)$ have the same meaning as in (8.1) for the object, K_o is resultant force on the the object, $v_o = \left[p^T,\theta^T\right]^T \in \mathbb{R}^n$ is the object frame's location and orientation vector $OX_OY_OZ_O$ in world frame $OX_WY_WZ_W$. Equation (8.10) fulfills the ensuing characteristics [24]:

Property 5

A positive definite symmetric matrix, $D_o\left(v_o\right)$, fulfills the following inequality:

$$\alpha_{D_o} I_n \le D_o\left(v_o\right) \le \beta_{D_o} I_n \tag{8.11}$$

where α_{D_o} and β_{D_o} consist of positive constants.

Property 6

The relation $\left(\dot{D}_o\left(v_o\right) - 2B_o\left(v_o,\dot{v}_o\right)\right)$ is skew symmetric.

8.3.2 Kinematics of multiple robot manipulators

Let $v_{oWB_i} = \left[p_{WB_i}^T,\theta_{WB_i}^T\right]^T$ is a position and orientation vector for base coordinate frame $OX_{B_i}Y_{B_i}Z_{B_i}$ w.r.t the world coordinate frame $OX_WY_WZ_W$ and $v_{oOE_i} = \left[p_{OE_i}^T,\theta_{OE_i}^T\right]^T$ is a position and orientation vector of end effector coordinate frame $OX_{E_i}Y_{E_i}Z_{E_i}$ with respect to the object coordinate frame $X_OY_OZ_O$. If $S_W^O, S_O^{E_i}, S_W^{B_i}$ and $S_{B_i}^{E_i}$ are the matrices of homogeneous transformation with expression

$S_W^O = \begin{bmatrix} \Re(\theta) & p \\ 0 & 1 \end{bmatrix}$, $S_O^{E_i} = \begin{bmatrix} \Re(\theta_{OE_i}) & p_{OE_i} \\ 0 & 1 \end{bmatrix}$, $S_W^{B_i} = \begin{bmatrix} \Re(\theta_{WB_i}) & p_{WB_i} \\ 0 & 1 \end{bmatrix}$, and

$S_{B_i}^{E_i} = \begin{bmatrix} \Re_i(q_i) & p_i(q_i) \\ 0 & 1 \end{bmatrix}$. Then these matrices satisfies the relation

$$S_W^O S_O^{E_i} = S_W^{B_i} S_{B_i}^{E_i} \tag{8.12}$$

or

$$S_W^O = S_W^{B_i} S_{B_i}^{E_i} \left(S_O^{E_i}\right)^{-1} \tag{8.13}$$

Equation (8.13) yields

$$p = p_{WB_i} + \mathfrak{R}\left(\theta_{WB_i}\right) p_i\left(q_i\right) - \mathfrak{R}\left(\theta_{WB_i}\right) \mathfrak{R}_i\left(q_i\right)\left[\mathfrak{R}\left(\theta_{WB_i}\right)\right]^T p_{OE_i} \tag{8.14}$$

$$\mathfrak{R}(\theta) = \mathfrak{R}\left(\theta_{WB_i}\right) \mathfrak{R}_i\left(q_i\right)\left[\mathfrak{R}\left(\theta_{OE_i}\right)\right]^T \tag{8.15}$$

Now Equation (8.14) and (8.15) can be revealed in compact form as

$$v_o = \left[p^T, \theta^T\right]^T = \left[\phi_{ri}^T\left(q_i\right), \phi_{\theta i}^T\left(q_i\right)\right]^T \tag{8.16}$$

Differentiation of (8.16) gives

$$\dot{v}_o = \mathfrak{E}_i \dot{q}_i, \; \ddot{v}_o = \mathfrak{E}_i \ddot{q}_i + \dot{\mathfrak{E}}_i \dot{q}_i \tag{8.17}$$

where $\mathfrak{E}_i = \left[\dfrac{\partial \phi_{ri}^T\left(q_i\right)}{\partial q_i}, \dfrac{\partial \phi_{\theta i}^T\left(q_i\right)}{\partial q_i}\right]^T$

8.4 OPTIMAL CONTROL LAW

We design the optimal control torque using Hamilton Jacobi Bellmen Optimization, RBF neural network and adaptive robust.

8.4.1 Integrated Dynamics of Object and Manipulators

Let $\mathfrak{E} = \left[\left(\mathfrak{E}_1^{-1}\right)^T, \cdots, \left(\mathfrak{E}_k^{-1}\right)^T\right]^T$ and $\mathfrak{H} = \left[\left(\mathfrak{E}_1^{-1}\dot{\mathfrak{E}}_1\mathfrak{E}_1^{-1}\right)^T, \cdots, \left(\mathfrak{E}_k^{-1}\dot{\mathfrak{E}}_k\mathfrak{E}_k^{-1}\right)^T\right]^T$. Then we have

$$\dot{q} = \mathfrak{E}\dot{v}_o, \ddot{q} = \mathfrak{E}\ddot{v}_o - \mathfrak{H}\dot{v}_o \tag{8.18}$$

Using Equation (8.18) and (8.3), we got the integrated dynamics of object and manipulator as

$$\begin{aligned} &D(q)\mathfrak{E}\ddot{v}_o + \left[B(q,\dot{q})\mathfrak{E} - D(q)\mathfrak{H}\right]\dot{v}_o + G(q) + F(\dot{q}) + T_d \\ &= \tau - J^T(q)\left[\mathfrak{Q}^+\left(D_o\left(v_o\right)\ddot{v}_o + B_o\left(v_o, \dot{v}_o\right)\dot{v}_o + G_o\left(v_o\right)\right) + \left(I - \mathfrak{Q}^+\mathfrak{Q}\right)\kappa_{int}\right] \end{aligned} \tag{8.19}$$

Multiplying this equation by $\mathfrak{C}^T$, it can be written as

$$\bar{D}\ddot{v}_o + \bar{B}\dot{v}_o + \bar{G} + \bar{F} + \bar{T}_d = \bar{\tau} - \bar{J}\kappa_{int} \tag{8.20}$$

where

$$\begin{aligned}
&\bar{D} = \mathfrak{C}^T D(q)\mathfrak{C} + \mathfrak{C}^T J^T(q)\mathfrak{Q}^+ D_o(v_o),\\
&\bar{B} = \mathfrak{C}^T B(q,\dot{q})\mathfrak{C} - \mathfrak{C}^T D(q)\mathfrak{H} + \mathfrak{C}^T J^T(q)\mathfrak{Q}^+ B_o(v_o,\dot{v}_o),\\
&\bar{G} = \mathfrak{C}^T G(q) + \mathfrak{C}^T J^T(q)\mathfrak{Q}^+ G_o(v_o),\\
&\bar{F} = \mathfrak{C}^T F(\dot{q}) = \mathfrak{C}^T F(\mathfrak{C}\dot{v}_o),\\
&\bar{T}_d = \mathfrak{C}^T T_d, \bar{\tau} = \mathfrak{C}^T \tau,\\
&\bar{J} = \mathfrak{C}^T J^T(q)(I - \mathfrak{Q}^+\mathfrak{Q}).
\end{aligned}$$

Equation (8.20) is integrated dynamics of object and manipulators. Moment/force at contact points is not incorporated by this dynamic model. Assumptions and properties that are utilized to make the system stable are as follows:

Property 7

The matrix $\bar{D}(v_o)$ is a positive definite, symmetric matrix that satisfies the inequality:

$$\alpha_{\bar{D}} I_n \le \bar{D}(v_o) \le \beta_{\bar{D}} I_n \tag{8.21}$$

where $\alpha_{\bar{D}}$ and $\beta_{\bar{D}}$ are positive constants.

Property 8

The relation $\left(\bar{D}(v_o) - 2\bar{B}(v_o,\dot{v}_o)\right)$ is skew symmetric.

Assumption 1

$\bar{F}(\dot{v}_o)\| \le c + d\|\dot{v}_o\|$ where $c \ge 0$ and $d \ge 0$,

Assumption 2

For positive constant γ, $\|\bar{T}_d\| \le \gamma$.

8.4.2 Error Dynamics

Assume that the desired trajectory of the internal force is and the desired trajectory of the moving object is κ_{intd}. Then, we define the position and orientation error as

$$e(t) = v_{od}(t) - v_o(t) \tag{8.22}$$

and

$$e_{\kappa}(t) = \tilde{\kappa}(t) + \int_{0}^{t} \tilde{\kappa}(t)\,dt \tag{8.23}$$

where $\tilde{\kappa}(t) = \kappa_{intd}(t) - \kappa_{int}(t)$ is the internal force error. Now define the filtered tracking error

$$r(t) = \dot{e}(t) + \Lambda e(t) \tag{8.24}$$

whereas Λ is symmetric positive definite matrix and

$$\eta(t) = \mathfrak{C} r(t) \tag{8.25}$$

If $r(t) = 0$ and $e_{\kappa}(t) = 0$, this implies $v_{od}(t) \to v_o(t)$ and $\tilde{\kappa}(t) \to 0$ as $t \to \infty$. Differentiating the Equation (8.24) and using Equation (8.22) in Equation (8.20), In relation to filtered tracking error, we have a system dynamics as

$$\bar{D}\dot{r}(t) = -\bar{B}r(t) - \bar{\tau} + g(y) \tag{8.26}$$

where $g(y) = \bar{D}(\ddot{v}_{od} + \Lambda\dot{e}) + \bar{B}(\dot{v}_{od} + \Lambda e) + \bar{G} + \bar{F} + \bar{T}_d + \bar{J}\kappa_{int}$ is a non linear function of unknown dynamics with $y(t) = \left[e^T, \dot{e}^T, v_{od}^T, \dot{v}_{od}^T, \ddot{v}_{od}^T\right]$.

Establish a control input in order to develop the best possible control input $Y(t)$ as

$$Y(t) = g(y) - \bar{\tau} \tag{8.27}$$

where $Y(t)$ is to be minimized later. Using (8.27) in (8.26), the dynamics becomes

$$\bar{D}\dot{r}(t) = -\bar{B}r(t) + Y(t) \tag{8.28}$$

Taking Equation (8.24) and Equation (8.28) together, we obtained the augmented form of system dynamics as

$$\dot{\tilde{v}}_o(t) = \begin{bmatrix} \dot{e} \\ \dot{r} \end{bmatrix} = \begin{bmatrix} -\Lambda & I \\ 0 & -\bar{D}^{-1}\bar{B} \end{bmatrix} \begin{bmatrix} e \\ r \end{bmatrix} + \begin{bmatrix} 0 \\ \bar{D}^{-1} \end{bmatrix} Y(t) \tag{8.29}$$

Or it can be expressed in state space error dynamics as

$$\dot{\tilde{v}}_o(t) = A(v_o, \dot{v}_o)\tilde{v}_o + C(v_o)Y(t) \tag{8.30}$$

where $A(v_o,\dot{v}_o)\in\mathbb{R}^{2nx2n}$, $C(v_o)\in\mathbb{R}^{2nxn}$, $\tilde{v}_o(t)=\begin{bmatrix}e^T & r^T\end{bmatrix}\in\mathbb{R}^{2n}$ is state variable and $Y(t)\in\mathbb{R}^n$ is control variable. Obtaining the best controller is the control objective $Y(t)$ in reference to the quadratic performance index $\beth(Y)$ is minimized subjected to (8.31).

8.4.2.1 Hamilton Jacobi Bellmen Optimization:

Define a quadratic performance index function $\beth(Y)$ as

$$\beth(Y)=\int_{t_0}^{\infty}\bar{L}(\tilde{v}_o,Y)\,dt \tag{8.31}$$

with Lagrangian functions

$$\begin{aligned}\bar{L}(\tilde{v}_o,Y)&=\frac{1}{2}\tilde{v}_o^T(t)H\tilde{v}_o(t)+\frac{1}{2}Y^T(t)MY(t)\\&=\frac{1}{2}\begin{bmatrix}e^T & r^T\end{bmatrix}\begin{bmatrix}H_{11} & H_{12}\\ H_{21} & H_{22}\end{bmatrix}\begin{bmatrix}e\\ r\end{bmatrix}+\frac{1}{2}Y^T(t)MY(t)\end{aligned} \tag{8.32}$$

Theorem 1

Let $Y^*(t)$ serve as the optimal control for achieving the control objective.

The necessary and sufficient condition are the $Y^*(t)$ to optimize the performnace index $\beth(Y)$ subjected to (8.30) is to select $Y_f=Y_f(\tilde{v}_o,t)$ that satisfies HJB [28]

$$\frac{\partial Y_f(\tilde{v}_o,t)}{\partial t}+\min_Y\left[Q\left(\tilde{v}_o,Y,\frac{\partial Y_f(\tilde{v}_o,t)}{\partial\tilde{v}_o},t\right)\right]=0 \tag{8.33}$$

When Q, the Hamiltonian function, is described as

$$Q\left(\tilde{v}_o,Y,\frac{\partial Y_f(\tilde{v}_o,t)}{\partial\tilde{v}_o},t\right)=\bar{L}(\tilde{v}_o,Y)+\frac{\partial Y_f(\tilde{v}_o,t)}{\partial\tilde{v}_o}\dot{\tilde{v}}_o \tag{8.34}$$

The function $Y_f(\tilde{v}_o,t)$ satisfies the following equation

$$-\frac{\partial Y_f(\tilde{v}_o,t)}{\partial t}=\bar{L}(\tilde{v}_o,Y^*)+\frac{\partial Y_f(\tilde{v}_o,t)}{\partial\tilde{v}_o}\dot{\tilde{v}}_o$$

and the minimum Hamiltonian obtained for optimal control $Y(t) = Y^*(t)$ is given by

$$Q^* = \min_{Y}\left[\bar{L}(\tilde{v}_o, Y) + \frac{\partial Y_f(\tilde{v}_o, t)}{\partial \tilde{v}_o}\dot{\tilde{v}}_o\right] = Q\left(\tilde{v}_o, Y^*, \frac{\partial Y_f(\tilde{v}_o, t)}{\partial \tilde{v}_o}, t\right) = -\frac{\partial Y_f(\tilde{v}_o, t)}{\partial t}$$

Lemma 1

If function Y_f is choosen as[28]

$$Y_f = \frac{1}{2}\tilde{v}_o^T R \tilde{v}_o = \frac{1}{2}\tilde{v}_o^T \begin{bmatrix} Z & 0 \\ 0 & \bar{D} \end{bmatrix} \tilde{v}_o \tag{8.35}$$

where $R = \begin{bmatrix} Z & 0 \\ 0 & \bar{D} \end{bmatrix}$, Z is positive symmetric matrix that satisfies the Riccati equation

$$\dot{R} + RA + A^T R + H - RCM^{-1}C^T R = 0 \tag{8.36}$$

with the following choice of relations

$$M^{-1} = H_{22} \tag{8.37}$$

$$Z = Z^T = -\frac{1}{2}\left(H_{12} + H_{12}^T\right) \tag{8.38}$$

and

$$\Lambda^T Z + Z\Lambda = H_{11} \tag{8.39}$$

Then, the optimal solution Y^* for Y is:

$$Y^*(t) = -M^{-1}C^T R\tilde{v}_o = -M^{-1}r(t) \tag{8.40}$$

With Y^*, the torque input $\overline{\tau}$ is computed as:

$$\overline{\tau} = g(y) - Y^*(t) \tag{8.41}$$

The system's stability is examined using the following theorem:

Theorem 2

If matrices Λ and Z satisfy (8.35) and R is bounded over the interval (t_0, ∞). Thus using $Y^*(t)$, the equation of controlled system (8.30) which is written as

$$\dot{\tilde{v}}_o(t) = \left\{A(v_o, \dot{v}_o) - C(v_o)M^{-1}C^T R\right\}\tilde{v}_o(t) \tag{8.42}$$

With respect to its origin $\mathbb{R}^{2n}$, is globally exponentially stable.

Proof: Since $Y_f(\tilde{v}_o,t)$ is increasing function & radially positive with $\|\tilde{v}_o\|$. Therefore, we select continous quadratic function $Y_f(\tilde{v}_o,t)$ as the Lyapunov candidate that has unique minimum. Now to prove globally exponential stablilty of *(8.42)* it is sufficient to prove that $\frac{dY_f}{dt}<0\ \forall$ $|\tilde{v}_o\|\neq 0$. Here, The solution of *(8.34) gives*

$$\frac{dY_f(\tilde{v}_o,t)}{dt}=-\bar{L}(\tilde{v}_o,Y^*) \tag{8.43}$$

Using (8.40) and (8.32) in (8.43), we get

$$\begin{aligned}\frac{dY_f(\tilde{v}_o,t)}{dt}&=-\frac{1}{2}\left\{\tilde{v}_o{}^T H\tilde{v}_o+\left(M^{-1}C^T R\tilde{v}_o\right)^T M\left(M^{-1}C^T R\tilde{v}_o\right)\right\}\\&=-\frac{1}{2}\left\{\tilde{v}_o{}^T H\tilde{v}_o+\left(C^T R\tilde{v}_o\right)^T\left(M^{-1}\right)^T MM^{-1}\left(C^T R\tilde{v}_o\right)\right\}\\&=-\frac{1}{2}\left\{\tilde{v}_o{}^T H\tilde{v}_o+\left(C^T R\tilde{v}_o\right)^T\left(M^{-1}\right)\left(C^T R\tilde{v}_o\right)\right\}<0,\ \forall\ t>0,\ \ \|\tilde{v}_o\|\neq 0\end{aligned} \tag{8.44}$$

Since $\frac{dY_f(\tilde{v}_o,t)}{dt}<0$, therefore Lyapunov 2nd theorem proves globally exponential stablility of the system (8.43)

8.4.3 Radial Basis Function Neural Network

An RBF is one specific kind of NN with a quick learning curve. There are two phases in the network training of the radial basis function. As seen in Figure 8.2, the weights are first calculated from the input layer to the hidden layer and then from the hidden layer to the output layer. It is used to estimate the nonlinear unknown dynamical system because of its clarity.

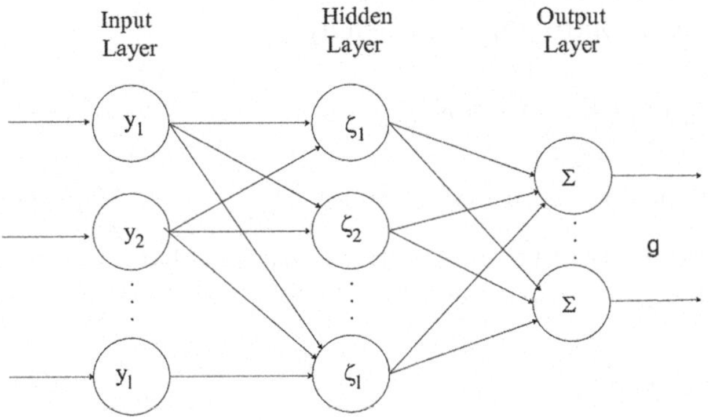

Figure 8.2 Radial basis function neural network.

Here is how to use RBFNN to estimate the unknown dynamics:

$$g(y) = W^T \zeta(y) + \epsilon(y) \tag{8.45}$$

where $W \in \mathbb{R}^{l \times n}$ indicates the weights of neural network, $\epsilon(y) : \mathbb{R}^{5n} \to \mathbb{R}^n$ represents the neural network reconstruction error and it is bounded by some ϵ_l and $\zeta(y) : \mathbb{R}^{5n} \to \mathbb{R}^l$ is a smooth basis function.

8.4.4 Adaptive bound

To define the adaptive bound β, considering assumptions 1 & 2 and upper bound ϵ_l for approximation error of neural network, we get

$$\left\| \overline{F} + \overline{T}_d + \epsilon(y) \right\| \leq c + d\left\| \dot{v}_o \right\| + \gamma + \epsilon_l = \beta \tag{8.46}$$

and β is of the following form

$$\beta = \begin{bmatrix} 1 & \left\| \dot{v}_o \right\| & 1 & 1 \end{bmatrix} \begin{bmatrix} c & d & \gamma & \epsilon_l \end{bmatrix}^T = \chi^T \left(\left\| \dot{v}_o \right\| \right) \Phi \tag{8.47}$$

where $\chi, \Phi \in \mathbb{R}^{p_1}$ are velocity and parameter vector respectively and p_1 is a fixed, positive number.

8.4.5 Optimal Controller Design Using RBFNN

NN and an adaptive compensator are used to design the ideal controller. Figure 8.2 depicts the proposed scheme's schematic diagram.

Let $\hat{g}(y)$ be functional approximation of $g(y)$, expressed as

$$\hat{g}(y) = \hat{W}^T \zeta(y) \tag{8.48}$$

where $\hat{W}$ is estimated NN weight.

The external control input is given by

$$\overline{\tau} = \hat{W}^T \zeta(y) - Y^* + \Delta + \mathfrak{C}^T K_s \eta \tag{8.49}$$

where K_s is positive gain matrix and $\Delta = \dfrac{\hat{\beta}^2 r}{\hat{\beta}\|r\| + \delta}$ is adaptive compensator with $\dot{\delta} = -\alpha\delta$, $\delta(0)$ and α are positive constant. Also $\hat{\beta} = \chi^T \hat{\Phi}$, whereas $\hat{\Phi}$ is the estimated value of parameter Φ. Placing (8.49) in (8.26), we attain

$$\overline{D}\dot{r}(t) = -\overline{B}r(t) - \left[\hat{W}^T \zeta(y) - Y^* + \frac{\hat{\beta}^2 r}{\hat{\beta}\|r\| + \delta} + \mathfrak{C}^T K_s \eta \right] + g(y) \tag{8.50}$$

$$\bar{D}\dot{r}(t) = -\bar{B}r(t) - \left[\hat{W}^T\zeta(y) - Y^* + \frac{\hat{\beta}^2 r}{\hat{\beta}\|r\| + \delta} + \mathfrak{E}^T K_s \eta\right] + W^T\zeta(y) + \bar{F} + \bar{T}_d + \epsilon(y) \tag{8.51}$$

$$\bar{D}\dot{r}(t) = -\bar{B}r(t) - \tilde{W}^T\zeta(y) + Y^* - \frac{\hat{\beta}^2 r}{\hat{\beta}\|r\| + \delta} - \mathfrak{E}^T K_s \eta + \bar{F} + \bar{T}_d + \epsilon(y) \tag{8.52}$$

Equation (8.52) in state space is provided by

$$\dot{\tilde{v}}_o(t) = A(v_o, \dot{v}_o)\tilde{v}_o + C(v_o) \left[Y^*(t) + \tilde{W}^T\zeta(y) - \frac{\hat{\beta}^2 r}{\hat{\beta}\|r\| + \delta} - \mathfrak{E}^T K_s \eta + \bar{F} + \bar{T}_d + \epsilon(y)\right] \tag{8.53}$$

where $\tilde{v}_o, A$ and C are as given in (8.30). Now, placing optimal control input in (8.53), w draw

$$\dot{\tilde{v}}_o(t) = \left(A - CM^{-1}C^T R\right)\tilde{v}_o + C\left(\tilde{W}^T\zeta(y) - \frac{\hat{\beta}^2 r}{\hat{\beta}\|r\| + \delta} - \mathfrak{E}^T K_s \eta + \bar{F} + \bar{T}_d + \epsilon(y)\right) \tag{8.54}$$

8.5 STABILITY EXPLORATION

Theorem 3

If the control input vector is given by (8.49) and error state space equation is given by (8.30), then error state $\tilde{v}_o$, Consequently, using the above adaptation law, the location and orientation tracking errors asymptotically converge to zero:

$$\dot{\hat{W}} = \daleth_W \zeta(y)\left(C^T R\tilde{v}_o\right)^T \tag{8.55}$$

$$\dot{\hat{\Phi}} = \daleth_\Phi \chi \|r\| \tag{8.56}$$

where $\daleth_W = \daleth_W^T \in \mathbb{R}^{l \times l}$ and $\daleth_\Phi = \daleth_\Phi^T \in \mathbb{R}^{p_1} \times p_1$ are the positive definite matrices.

Proof: Examine the following Lyapunov function.

$$L_f = \frac{1}{2}\tilde{v}_o^T R\tilde{v}_o + \frac{1}{2}tr\left(\tilde{W}^T \daleth_W^{-1}\tilde{W}\right) + \frac{1}{2}tr\left(\tilde{\Phi}^T \daleth_\Phi^{-1}\tilde{\Phi}\right) + \frac{\delta}{\alpha} \tag{8.57}$$

Taking first derivative of function L_f, we have

$$\dot{L}_f = \tilde{v}_o^T R \dot{\tilde{v}}_o + \frac{1}{2}\tilde{v}_o^T \dot{R} \tilde{v}_o + tr\left(\tilde{W}^T \daleth_W^{-1} \dot{\tilde{W}}\right) + tr\left(\tilde{\Phi}^T \daleth_\Phi^{-1} \dot{\tilde{\Phi}}\right) + \frac{\dot{\delta}}{\alpha} \tag{8.58}$$

Substituting (8.54) in (8.58) and using the relation $\dot{\tilde{W}} = -\dot{\hat{W}}, \dot{\tilde{\Phi}} = -\dot{\hat{\Phi}}$ and $\dot{\delta} = -\alpha\delta$, we get

$$\dot{L}_f = \tilde{v}_o^T R \left[\left(A - CM^{-1}C^T R\right)\tilde{v}_o + C\left(\tilde{W}^T \zeta(y) - \frac{\hat{\beta}^2 r}{\hat{\beta}\|r\| + \delta} + \bar{F} + \bar{T}_d + \epsilon(y)\right)\right]$$
$$+ \frac{1}{2}\tilde{v}_o^T \dot{R} \tilde{v}_o - tr\left(\tilde{W}^T \daleth_W^{-1} \dot{\hat{W}}\right) - tr\left(\tilde{\Phi}^T \daleth_\Phi^{-1} \dot{\hat{\Phi}}\right) - \delta \tag{8.59}$$

$$= \tilde{v}_o^T \left(RA + \frac{1}{2}\dot{R}\right)\tilde{v}_o - \tilde{v}_o^T RCM^{-1}C^T R\tilde{v}_o + \tilde{v}_o^T RC\tilde{W}^T \zeta(y) - \tilde{v}_o^T RC \frac{\hat{\beta}^2 r}{\hat{\beta}\|r\| + \delta}$$

$$+ \tilde{v}_o^T RC\left[\bar{F} + \bar{T}_d + \epsilon(y)\right] - tr\left(\tilde{W}^T \daleth_W^{-1} \dot{\hat{W}}\right) - tr\left(\tilde{\Phi}^T \daleth_\Phi^{-1} \dot{\hat{\Phi}}\right) - \delta \tag{8.60}$$

Now

$$\text{RA} = \frac{1}{2}\left\{A^T R + RA\right\} \tag{8.61}$$

and from Riccati Equation (8.36), we get

$$\frac{1}{2}A^T R + \frac{1}{2}RA + \frac{1}{2}\dot{R} = -\frac{1}{2}H + \frac{1}{2}RCM^{-1}C^T R^T \tag{8.62}$$

Using (8.61) and (8.62) in (8.60), we have

$$\dot{L}_f = \tilde{v}_o^T \left(-\frac{1}{2}H + \frac{1}{2}RCM^{-1}C^T R^T\right)\tilde{v}_o - \tilde{v}_o^T RCM^{-1}C^T R\tilde{v}_o$$
$$+ \tilde{v}_o^T RC\tilde{W}^T \zeta(y) - \tilde{v}_o^T RC\frac{\hat{\beta}^2 r}{\hat{\beta}\|r\| + \delta} + \tilde{v}_o^T RC\left[\bar{F} + \bar{T}_d + \epsilon(y)\right]$$
$$- tr\left(\tilde{W}^T \daleth_W^{-1} \dot{\hat{W}}\right) - tr\left(\tilde{\Phi}^T \daleth_\Phi^{-1} \dot{\hat{\Phi}}\right) - \delta = -\frac{1}{2}\tilde{v}_o^T \left(H + RCM^{-1}C^T R^T\right)\tilde{v}_o$$
$$+ \tilde{v}_o^T RC\tilde{W}^T \zeta(y) - \tilde{v}_o^T RC\frac{\hat{\beta}^2 r}{\hat{\beta}\|r\| + \delta} + \tilde{v}_o^T RC\left[\bar{F} + \bar{T}_d + \epsilon(y)\right]$$
$$- tr\left(\tilde{W}^T \daleth_W^{-1} \dot{\hat{W}}\right) - tr\left(\tilde{\Phi}^T \daleth_\Phi^{-1} \dot{\hat{\Phi}}\right) - \delta \tag{8.63}$$

If the adaptation laws given by (8.55) and (8.56) are used, then above equation can be rewritten as

$$\dot{L}_f = -\frac{1}{2}\tilde{v}_o^T U \tilde{v}_o + \tilde{v}_o^T RC\tilde{W}^T \zeta(y) - \tilde{v}_o^T RC \frac{\hat{\beta}^2 r}{\hat{\beta}\|r\| + \delta} + \tilde{v}_o^T RC\left[\bar{F} + \bar{T}_d + \epsilon(y)\right]$$
$$- tr(\tilde{W}^T \zeta(y)\left(C^T R\tilde{v}_o^T\right) - tr\left(\tilde{\Phi}^T \chi\|r\|\right) - \delta = -\frac{1}{2}\tilde{v}_o^T U \tilde{v}_o - \tilde{v}_o^T RC \frac{\hat{\beta}^2 r}{\hat{\beta}\|r\| + \delta}$$
$$+ \tilde{v}_o^T RC\left[\bar{F} + \bar{T}_d + \epsilon(y)\right] - tr\left(\tilde{\Phi}^T \chi\|r\|\right) - \delta \tag{8.64}$$

where $U = H + RCM^{-1}C^T R^T$. Since $RC = \begin{bmatrix} 0 & I \end{bmatrix}^T$ and employ the value of adaptive bound β, we attain

$$\left\|\tilde{v}_o^T RC\left[\bar{F} + \bar{T}_d + \epsilon(y)\right]\right\| = \left\|\tilde{v}_o^T RC\right\|\left\|\bar{F} + \bar{T}_d + \epsilon(y)\right\|$$
$$= \|r\|\left\|\bar{F} + \bar{T}_d + \epsilon(y)\right\| \leq \chi^T \Phi \|r\| = \chi^T\left(\hat{\Phi} + \tilde{\Phi}\right)\|r\| \tag{8.65}$$

Now putting inequality (8.65) in (8.64), we get

$$\dot{L}_f \leq -\frac{1}{2}\tilde{v}_o^T U \tilde{v}_o - \frac{\left(\chi^T \hat{\Phi}\right)^2 \|r\|^2}{\chi^T \hat{\Phi}\|r\| + \delta} + \chi^T\left(\hat{\Phi} + \tilde{\Phi}\right)\|r\| - tr\left(\tilde{\Phi}^T \chi\|r\|\right) - \delta$$
$$\leq -\frac{1}{2}\tilde{v}_o^T U \tilde{v}_o - \frac{\left(\chi^T \hat{\Phi}\right)^2 \|r\|^2}{\chi^T \hat{\Phi}\|r\| + \delta} + \chi^T \hat{\Phi}\|r\| - \delta$$
$$\leq -\frac{1}{2}\tilde{v}_o^T U \tilde{v}_o - \frac{\delta^2}{\chi^T \hat{\Phi}\|r\| + \delta} \tag{8.66}$$

This implies

$$\dot{L}_f \leq -\frac{1}{2}\sigma_{\min}\|\tilde{v}_o\|^2 \leq 0 \tag{8.67}$$

where $\sigma_{\min}$ is the matrix's minimal single value U. So, $L_f > 0$ & $\dot{L}_f \leq 0$, this implicit that $\tilde{v}_o, \tilde{W}, \tilde{\Phi}$ & hence the $\hat{\Phi}, \hat{W}$ are bounded. If $\Xi(t)$ is an expression of a function by

$$\Xi(t) = \sigma_{\min}\|\tilde{v}_o\|^2 \tag{8.68}$$

Then, $\Xi(t) \leq -\dot{L}_f$. Integration of Equation (8.68) gives the relation as

$$\int_0^t \Xi(t)dt \leq L_f(0) - L_f(t) \tag{8.69}$$

we have

$$\lim_{t\to\infty}\int_0^t \Xi(t)\,dt < \infty \tag{8.70}$$

Equation (8.70) exists as $L_f(t)$ is bounded and decreasing function. Also $\tilde{v}_o(t)$ is bounded so $\dot{\tilde{v}}_o(t)$ is bounded and this implies $\dot{\Xi}(t)$ is bounded. It indicates uniformly continuity of $\Xi(t)$. Then $\lim_{t\to\infty}\Xi(t)=0 \Rightarrow \tilde{v}_o(t)\to 0$ as $t\to\infty$ (Barbalat's lemma [31]). This gives error $e(t)\to 0$ as $t\to\infty$. It shows that the errors are converging asymptotically and system is stable.

8.6 NUMERICAL SIMULATION

The suggested method is used on two three-link planar robot manipulators that are firmly holding a circular disk. The simulation results are presented in comparison manner for force/position tracking, joint tracking of robot manipulators. The details of model dynamics can be found in [24].

The object dynamics model is taken as

$$\begin{bmatrix} b_0 & 0 & 0 \\ 0 & b_0 & 0 \\ 0 & 0 & I_0 \end{bmatrix}\begin{bmatrix} \ddot{x} \\ \ddot{y} \\ \ddot{\theta} \end{bmatrix}+\begin{bmatrix} 0 \\ b_0 g \\ 0 \end{bmatrix}=\begin{bmatrix} \kappa_{0x} \\ \kappa_{0y} \\ \mu_0 \end{bmatrix}$$

where (x,y) is the centre of the circular disk with respect to the world frame and θ is the angle of diameter. Kinematic relationship of two robot manipulators is expressed with vectors $p_{WB_1}=\left[-h_1,0,0\right]^T$, $p_{WB_2}=\left[-h_2,0,0\right]^T$, $p_{OE_1}=\left[-p,0,0\right]^T$, $p_{OE_2}=\left[p,0,0\right]^T$ and the object position and orientation related with joint space of manipulators are presented as:

$$x=-h_1+l_{11}\cos(q_{11})+l_{12}\cos(q_{11}+q_{12})+(p+l_{13})\cos(q_{11}+q_{12}+q_{13});$$

$$=h_2+l_{21}\cos(q_{21})+l_{22}\cos(q_{21}+q_{22})+(p+l_{23})\cos(q_{21}+q_{22}+q_{23});$$

$$y=l_{11}\sin(q_{11})+l_{12}\sin(q_{11}+q_{12})+(p+l_{13})\sin(q_{11}+q_{12}+q_{13});$$

$$=l_{21}\sin(q_{21})+l_{22}\sin(q_{21}+q_{22})+(p+l_{23})\sin(q_{21}+q_{22}+q_{23});$$

$$\theta=q_{11}+q_{12}+q_{13};$$

$$=q_{21}+q_{22}+q_{23};$$

The parameter value for the manipulators are given as $h_1 = 1, h_2 = 1, l_{11} = l_{12} = l_{21} = l_{22} = 2.05, l_{13} = l_{23} = 0.5, b_{11} = b_{12} = b_{21} = b_{22} = 1, b_{13} = b_{23} = 0.3, b_0 = 0.2, I_0 = 0.1$, g = 9.8. The performance index uses weighting matrices, which are defined as follows: $H_{11} = 10I_2$, $H_{12} = -\begin{bmatrix} 0.5 & 0 \\ 0 & 5 \end{bmatrix}, H_{21} = H_{12}^T, H_{22} = M^{-1} = 40I_2$. Solving for matrices Z and Λ, we have, $Z = \begin{bmatrix} 0.5 & 0 \\ 0 & 5 \end{bmatrix}$ and $\Lambda = \begin{bmatrix} 10 & 0 \\ 0 & 1 \end{bmatrix}$. The desired trajectory for circular disk is given as $x_d(t) = 0.5\cos(\lambda t)$; $y_d(t) = 2 + 0.5\sin(\lambda t)$; $\theta(t) = 0$ where $\lambda = 0.5\pi$. The desired internal force is $\kappa_{intd} = 0$. For the architecture of RBFNN we take 10 nodes with centre $c_i \in [-2, 2]^{10}$ and $\sigma_i = 0.2$. The update law weights are $\urcorner_W = 10I_{10}$, $\urcorner_\Phi = 10I_4$ and gain matrix $K_s = 20I$.

In order to demonstrate the efficacy of the suggested controller, we conducted simulation simulations using the following scenarios:

Case I: In this case, we use neural network based controller [24].
Case II: Here, we use proposed controller.

Now the performance of above defined various controllers are shown by Figures 8.3–8.5. Figure Figures 8.3(a), 8.3(b) shows the performance of case I controller. From these figures, it is observed that the position and orientation errors are small and system is uniformly ultimate bounded i.e. the current controller makes an effort to reduce system uncertainty, and we receive satisfactory replies from it. Further when we move on the results of proposed controller as in Case II, which are produced in Figures 8.4, Figure 8.4(a), position and orientation errors, Figure 8.4(b), joint tracking error, shows the performance the proposed controller and it is evident from these figures that the controller can effectively overcome the structured and unstructured uncertainties in the system as compared to other controllers and errors converging to zero level asymptotically. Also, the internal force trajectory tracking can be seen in Figure 8.5 which are essential for the purpose of proper contact with the object.

In order to quantitatively compare the controllers' performance, a scaler valued L^2 norm [24] is used for position and orientation error curves which calculate the RMS average of errors. For tracking error, the L^2 norm is defined as follows:

$$L^2[e] = \sqrt{\frac{1}{t_f - t_0} \int_{t_0}^{t_f} \|e(t)\|^2 \, dt} \tag{8.71}$$

Whereas t_0 is initial time and t_f is final timeThe control strategy performs better when the RMS average is modest. Table 8.1 demonstrates how

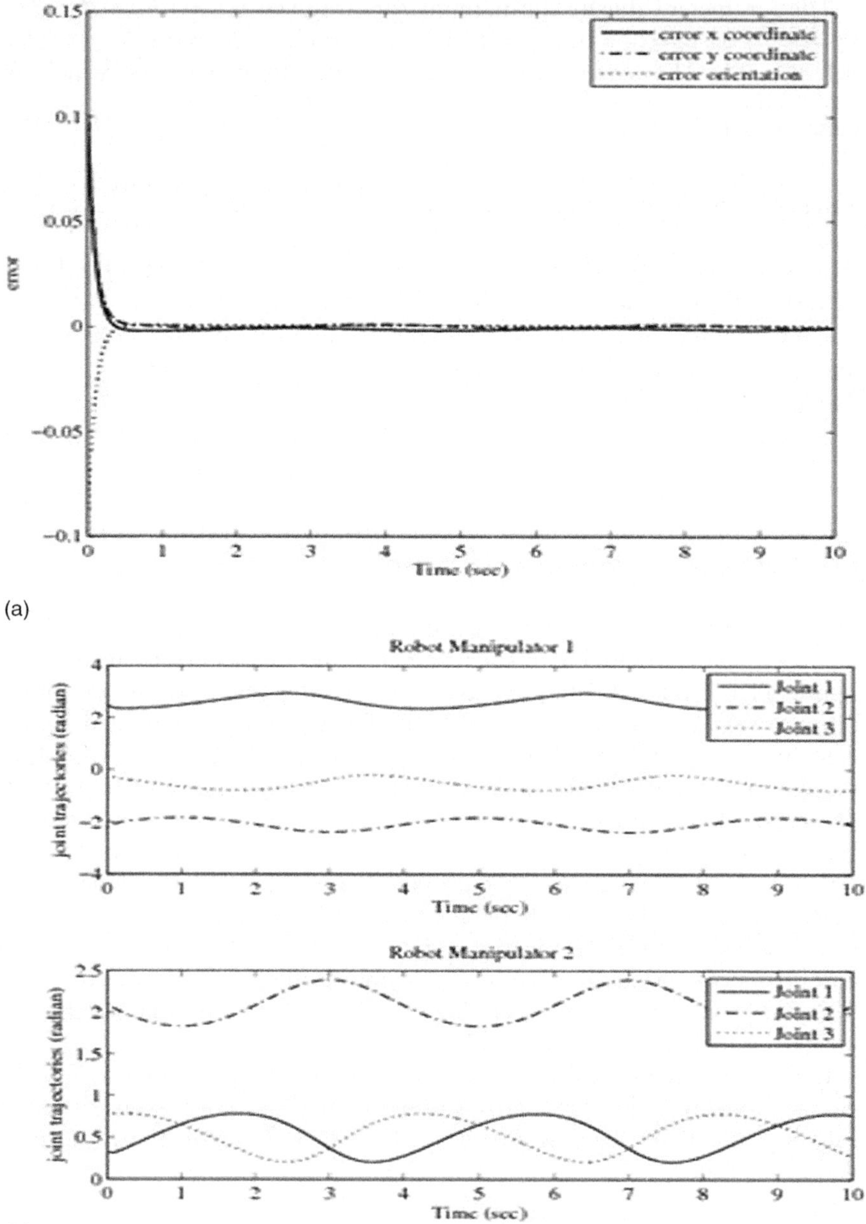

Figure 8.3 Performance of controller for Case I: (a) position and orientation error (b) joint tracking of robot manipulators.

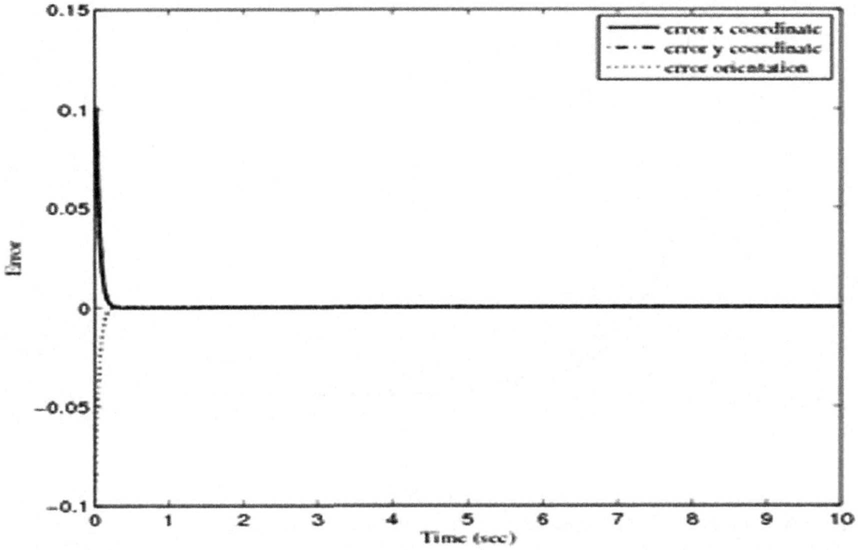

(a)

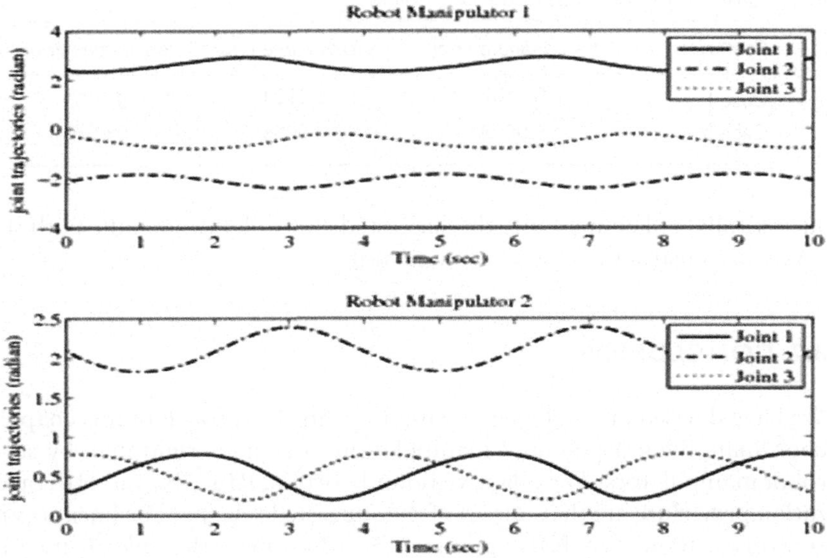

(b)

Figure 8.4 Performance of controller for Case II: (a) position and orientation error (b) joint tracking of robot manipulators.

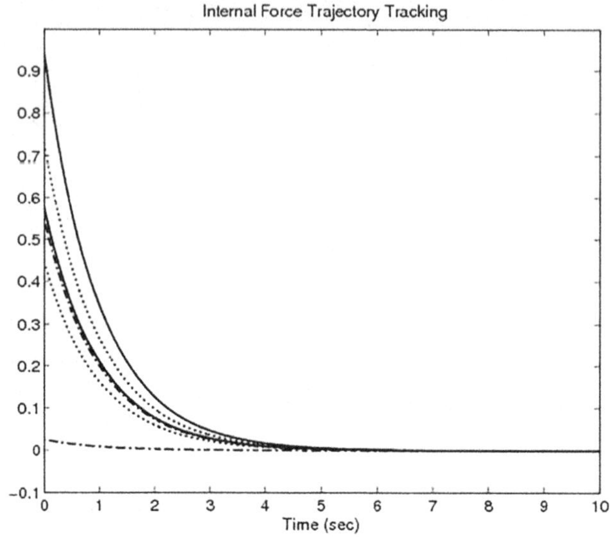

Figure 8.5 Internal force tracking with proposed controller.

Table 8.1 RMS average of tracking errors

Controllers	*x-coordinate error*	*y-coordinate error*	*orientation error*
Case I	0.0102	0.0103	0.1000
Case II	0.0098	0.0093	0.0079

successfully and comparably the suggested controller can handle both structured and unstructured system uncertainty.

8.7 CONCLUSION

We have devised an intelligent optimum control approach in this chapter to coordinate the grasping and manipulation of a single rigid item by several robot manipulators. The adaptive bound portion, RBFNN, and HJB optimization procedure are the sources of the suggested scheme. The linear optimal control is attained by HJB optimization. To estimate the unknown dynamics, RBFNN is used. To approximate the unknown boundaries on the NN estimation error and system uncertainties, the adaptive compensator is used. The asymptotic stability of the system is analyzed using Lyapunov stability and optimal control theory. The suggested controller adjusts for a variety of system uncertainties. Finally, a comparative analysis of the simulation portion demonstrates the productivity of the suggested control strategy.

REFERENCES

[1] Alford C, Belyeu S. Coordinated control of two robot arms. *Proceedings of the IEEE Int. Conf. Robotics and Automation*, 1984: 468–473.

[2] Arimoto S, Miyazaki F, Kawamura S. Cooperative motion control of multiple robot arms or fingers. *Proceedings of the IEEE Int. Conf. Robotics and Automation*, 1987: 1407–1412.

[3] Hayati S. Hybrid position/force control of multi-arm cooperating robots. *Proceedings of the IEEE Int. Conf. Robotics and Automation*, 1986: 82–89.

[4] Tarn T, Bejczy A, Yun X. Design of dynamic control of two cooperating robot arms: Closed chain formulation. *Proceedings of the IEEE Int. Conf. Robotics and Automation*, 1986: 7–13.

[5] Uchiyama M, Dauchez P. A symmetric hybrid position/force control scheme for the coordination of two robots. *Proceedings of the IEEE Int. Conf. Robotics and Automation*, 1988: 350–356.

[6] Ping H. Coordinated control of multiple manipulator systems. *IEEE Transactions on Robotics and Automation*, 1993, 9(4): 400–410.

[7] Yun X, Kumar V. An approach to simultaneous control of trajectory and interaction forces in dual-arm configurations. *IEEE Transactions on Robotics and Automation*, 1991, 7(5), 618–625.

[8] Ning X, Tarn T, Bejczy A. Intelligent planning and control for multirobot coordination: An event-based approach. *IEEE Transactions on Robotics and Automation*, 1996,12(3): 439–452.

[9] Hu Y R, Goldenberg A. An adaptive approach to motion and force control of multiple coordinated robot arms. *Proceedings of the IEEE Int. Conf. Robotics and Automation*, 1989: 1091–1096.

[10] Walker J D M, Kim D. Adaptive coordinated motion control of two manipulator arms. *Proceedings of the IEEE Int. Conf. Robotics and Automation*, 1989: 1084–1090.

[11] Yao B, Gao W B, Chan S P, Cheng M. coordinated control of two manipulator arms in the presence of environmental constraints. *IEEE Transactions on Automatic Control*, 1982, 37(11): 1806–1812.

[12] Jean L F J An adaptive control scheme for coordinated multimanipulator systems. *IEEE Transactions on Robotics and Automation*, 1993, 9: 226–231.

[13] Gueaieba M B W, Sharhanb S. Robust computationally efficient control of cooperative closed-chain manipulators with uncertain dynamics. *Automatica*, 2007, 43: 842–851.

[14] Hong J S Z W Control of two industrial manipulators rigidly holding an egg. *IEEE Control System Magazine*, 1999, 19(2): 24–30.

[15] Namvar, M, Aghili F. Adaptive force-motion control of coordinated robots interacting with geometrically unknown environments. *IEEE Transaction Robotics*, 2005, 21(4): 678–694.

[16] Trinh L, Mikael E, Baran C. Dependable navigation for multiple autonomous robots with petri nets based congestion control and dynamic obstacle avoidance. *Journal of Intelligent & Robotic Systems*, 2022. https://doi.org/10.1007/s10846-022-01589-1

[17] Rani K, Kumar N. Design of intelligent hybrid force and position control of robot manipulator. *Procedia Computer Science*, 2018, 125: 42–49.

[18] Rani K, Kumar N. Intelligent controller for hybrid force and position control of robot manipulators using RBF neural network. *International Journal of Dynamics and Control*, 2019, 7(2): 767–775.

[19] Bouazza H, Bendaas M L, Allaoui T, Denai M. Application of artificial intelligence to wind power generation: Modelling, control and fault detection. *International. Journal of Intelligent Systems Technologies and Applications*, 2020, 19(3): 280–305. Springer Nature 2021 LATEX template.

[20] Cui R, Yan W. Mutual synchronization of multiple robot manipulators with unknown dynamics. *Journal of Intelligent & Robotic Systems*, 2012, 68(2): 105–119.

[21] Yu L, Fei S, Sun L, Huang J, Yang G. Design of robust adaptive neural switching controller for robotic manipulators with uncertainty and disturbances. *Journal of Intelligent & Robotic Systems*, 2015, 77(3):571–581.

[22] Hou M T Z, Cheng L. Coordination of two redundant robots using a dual neural network. *Proceedings of the International Joint Conference on Neural Network*, 2006, 4187–4192.

[23] Siqueira M T A Neural network-based control for fully actuated and under-actuated cooperative manipulators. *Control Engineering Practice*, 2009, 17: 418–425.

[24] Panwar V, Kumar N, Borm N S J. Adaptive neural controller for cooperative multiple robot manipulator system manipulating a single rigid object. *Applied Soft Computing*, 2012, 12: 216–227.

[25] Rani M, Kumar N. A new hybrid position/- force control scheme for coordinated multiple mobile manipulators. *Arabian Journal for Science and Engineering*, 2019, 44(3): 2399–2411.

[26] Zhang S, Lei M, Dong Y, He W. Adaptive neural network control of coordinated robotic manipulators with output constraint. *IET Control Theory & Applications*, 2016, 10(17): 2271–2278.

[27] Kim Y, Lewis F, Dawson D. Intelligent optimal control of robotic manipulators using neural networks. *Automatica*, 2000, 36, 1355–1364.

[28] Panwar V, Sukavanam N. Design of optimal hybrid position/force controller for a robot manipulator using neural network. *Mathematical Problems in Engineering*, 2007, 23: 1–23.

[29] Dlimi I B, Kallel H. Optimal neural control for constrained robotic manipulators. *Proceedings of the 5th IEEE International Conference in Intelligent Systems (IS)*, London, UK, 2010: 302–308.

[30] Ornelas-Tellez F, Sanchez E, Loukianov A. Discrete time neural inverse optimal control for nonlinear systems via passivation. *IEEE Transactions on Neural Networks and Learning Systems*, 2012, 23(8):1327–1339.

[31] Mehraeen S, Jagannathan S. Decentralized optimal control of a class of interconnected nonlinear discrete-time systems by using online Hamilton-Jacobi-Bellman formulation. *IEEE Transactions on Neural Networks*, 2011, 22(11): 1757–1769.

[32] Mannava A, Balakrishnan S, Tang T, Landers R. Optimal tracking control of motion systems. *IEEE Transactions on Control Systems Technology*, 2012, 20(6): 1548–1558.

[33] Sadati, N., & Babazadeh, A. A new approach for optimal control of multiple-arm robotic systems. *IFAC Proceedings*, *38*(1), 313–318, 2005.

[34] Rani K, Kumar N. An optimal control approach for hybrid motion/force control of coordinated multiple nonholonomic mobile manipulators using neural network. *International Journal of Modelling, Identification and Control*, 2021, 37(2), 164–175.
[35] Li Y, Chen L, Tee K, Li Q. Reinforcement learning control for coordinated manipulation of multi-robots. *Neurocomputing*, 2015, 170: 168–175.
[36] Sieber D, Deroo F, Hirche S. Formation-based approach for cooperative manipulation based on iterative optimal controller design. *Proc. IEEE/RSJ International Conference on Intelligent Robots and Systems (IROS)*, 2013.

Chapter 9

Emerging trends of machine learning and Internet of Things-based applications in day-to-day life

Pummy Dhiman, Sana Bharti and Bhavna Sharma

9.1 INTRODUCTION

Data science, which combines various scientific fields, including data mining and machine learning (ML), will play a major role in enhancing the intelligence of Internet of Things (IoT) applications [1]. Modern cloud apps that are heavily reliant on the IoT and ML are integral parts of our daily lives. Around 25 billion devices and gadgets, from wearables and automated cars to smart homes and smart city apps, are interacting with the IoT, according to a recent report by Gartner [2]. The IoT is expected to be used widely; by 2030, it is predicted that 29.42 billion "things" will be online. To ensure that these networked (smart) devices continuously learn from the already available data sets and improve themselves without human intervention, a vast amount of data generated by all of these devices must be analyzed. This is the point at which ML prerequisites emerge. The fusion of ML algorithms with IoT technology is revolutionizing smart homes, wearables [3], urban infrastructure [4], and healthcare systems [5]. Understanding the consequences of this convergence is imperative, given the growing prevalence of these technologies that are transforming our methods of work, communication, and environmental awareness. Without incorporating and enforcing ML, it would be extremely difficult for smart devices to gradually make intelligent decisions. In order to collect and distribute data, the IoT assists in connecting a variety of hardware items, including homes, vehicles, electronic devices, and other devices that are combined with gateways, sensors, and software. As more businesses become aware of the IoT' potential for advancement, they are starting to identify the obstacles that need to be removed in order to use it effectively. Many companies and organizations use ML to take advantage of the latent potential of IoT.

This chapter assesses the various ML techniques that address the difficulties associated with managing IoT data. Smart devices generate big data and communicate with each other and the data is sent to the cloud for analysis and storage, this concept is known as IoT. Further sub-sections, the concepts of IoT and ML are explored.

DOI: 10.1201/9781003590439-9

9.1.1 Internet of Things

Internet of Things refers to a vast network of devices that are embedded with sensors, software, and various other mechanisms that actively collect and exchange data over the internet [6]. These devices range from everyday objects like wearables to more extensive industrial machinery to infrastructure components. This interconnection allows a seamless flow of information and enhances the efficiency and functionality of the devices involved.

The main components are:

1. **Sensors and Actuators**: IoT devices incorporated with built to actively collect information about their surroundings like location, motion, temperature, and humidity. These devices are equipped mechanism with actuators allow these devices to take actions, such as regulating temperature, sounding alarms, or operating machinery, based on the data they gather [7].
2. **Connectivity**: IoT device employ a variety of communication protocols to send data to other devices or enteral To send data to other devices or central systems They can utilize both wired and wireless networks, cellular networks, Wi-Fi, Bluetooth, Zigbee, and LoRaWAN, to establish this connectivity. Through these diverse networks, IoT ensure seamless data transfer and enhancing their functionality and integration into various environment from smart homes to industrial settings.
3. **Data Processing and Analysis**: In order to obtain valuable insights, identify trends, and make data-driven decisions we process and examine the data gathered by IoT devices. This process might involve cloud computing which involves sending data to remove servers for processing. Alternatively, it might utilize edge computing where data undergoes processing locally on the device itself. By doing so we can leverage the vast amount of data collected from various sources.
4. **Integration with Applications and Services**: IoT systems frequently connect with services and software programs providing value added functionality. This functionality might include web-based dashboards designed for the detailed visualization of data, smart phone apps that allow users to remotely control and monitor devices, or API design to integrate IoT data seamlessly with existing enterprise systems. These systems enable more efficient operation by offering user real time data.

Figure 9.1 represents a typical IoT architecture, including:

1. **Devices**: These are the actual physical objects that interact with the surroundings and gather data thanks to sensors and actuators.
2. **Gateway**: Combining data from several devices and sending it for additional processing acts as a bridge between devices and the rest of the IoT system.

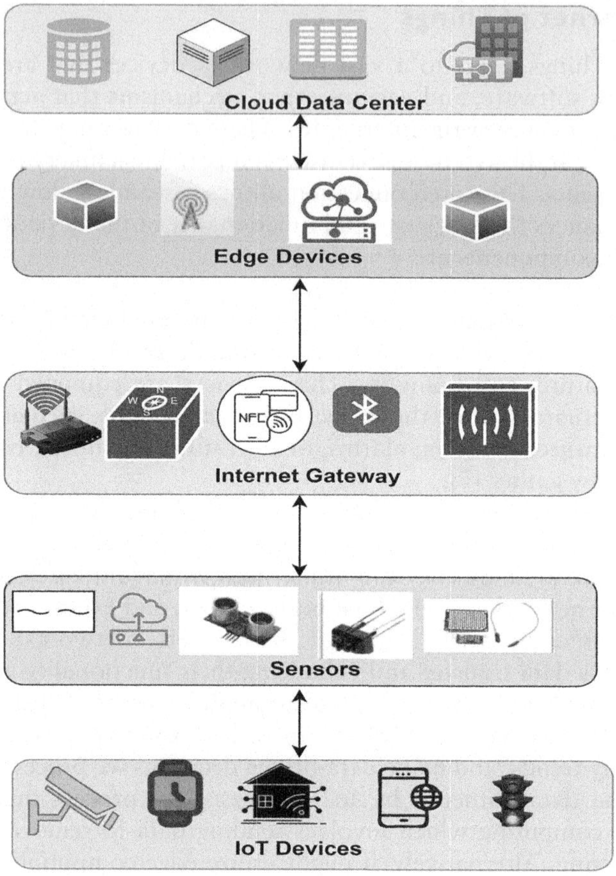

Figure 9.1 Architecture of IoT.

3. **Edge Computing**: This layer filters and aggregates data before sending it to the cloud, performing data processing and analysis closer to the source (i.e., devices), minimizing latency and bandwidth utilization [8].
4. **Cloud Services**: Here, data from edge devices is further processed, saved, and examined using cloud-based services. Data storage, ML techniques, and application interfaces are a few examples of these services.

9.1.2 Machine learning

Machine learning is a subfield of artificial intelligence and computer science that strives to use data and algorithms to learn more accurately in a way that is similar to how people learn [9]. In order to train algorithms to make predictions or classifications and to discover important insights in

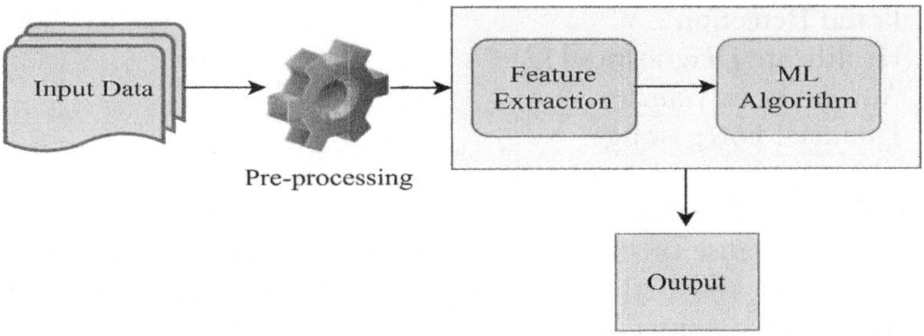

Figure 9.2 Machine learning framework.

data mining tasks, statistical methods are used. It is common practice to employ ML algorithms for the purposes of classification and prediction. These algorithms take in data, either labeled or unlabeled, and use it to make predictions about what the data pattern might be. This model's prediction is assessed using an error function. ML models are classified into three types [10]: supervised learning, which uses labeled datasets to train algorithms; unsupervised learning, which analyzes and clusters unlabeled datasets without human intervention; and semi-supervised learning, which uses a smaller, labeled data set to facilitate classification and feature extraction from a larger, unlabeled data set during training. The learning process of a ML algorithm for gaining new information utilizing existing knowledge and behavior is depicted in Figure 9.2.

Key components of ML are:

1. Data: which can be semi-structured, unstructured, or structured.
2. Algorithms, which find patterns and relationships in data
3. Model Training: In order to train a ML model to effectively predict outcomes or execute tasks on unseen data, labeled data (input-output pairs) are fed to the model, and its parameters are adjusted iteratively [10].
4. Testing and Evaluation: After training, ML models must be tested on untested data to measure their effectiveness and capacity for generalization. The results of testing aid in determining how effectively the model will function in actual situations [11].
5. Deployment and Monitoring: In order to deploy a ML model in a production setting where it can generate predictions or carry out tasks instantly, it must be integrated. To guarantee that the model continues to operate at its best and to identify any drift or degradation over time, continuous monitoring is necessary.

Machine learning can be used in various fields, like

1. Image Speech Recognition
2. Natural language processing [11]

3. Fraud Detection
4. Healthcare Diagnostics [12]
5. Autonomous Vehicles
6. Financial Forecasting

ML is still transforming businesses and encouraging innovation in many other fields. Because of its capacity to forecast, derive insights from large volumes of data, and analyze them all, it is a vital tool for both scholars and enterprises. The potential for ML to tackle difficult problems and enhance decision-making will only increase as more data becomes available and technology develops.

9.1.3 The intersection of machine learning and internet of things

Machine learning and the IoT have a strong connection that is changing several applications and industries. Sensors are used in IoT systems to gather much data from the physical world, including motion, sound, and temperature, and humidity. In the meantime, ML methods enable systems to learn from this data and make wise decisions with minimal human intervention. Actuators then receive control signals from applied ML algorithm and perform the desired actions. The intersection phenomena are shown in Figure 9.3. ML algorithms, have the capability to analyze sensor data in smart home applications, which enables to detect patterns in tenant behavior and subsequently optimize energy consumption. For example by evaluating the way tenants use lights and heating systems, these algorithms can adjust settings in real time to save energy without compromising comfort. Similarly in an industrial context analyzing sensor data from machinery allows ML- powered predicted maintenance systems to anticipate issues before they occur. This proactive approach helps avoid downtime and expensive repairs, ensuring continuous operations. By integrating

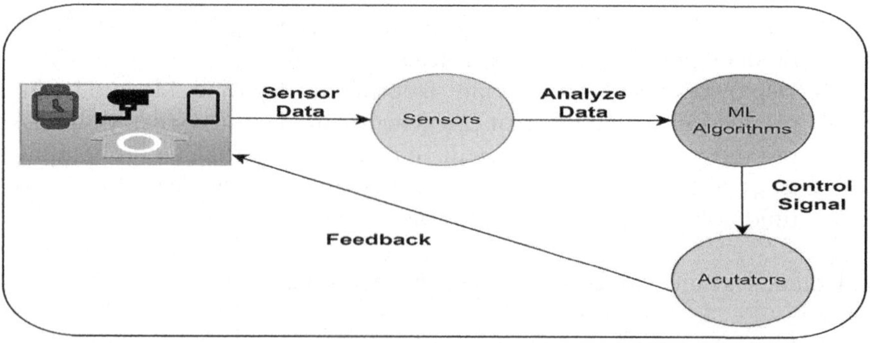

Figure 9.3 Intersection of ML and IoT.

ML with IoT systems can automate decisions and actions, enhancing both user experience and operational efficiency. Thai combination paves the way for intelligence, where devices can learn from patterns and make informed decisions, leading to more responsive and efficient environment [4].

9.1.3.1 Contribution

- This study examines the growing influence of ML and IoT-based applications in daily life, highlighting their impact across sectors as well as their impact on homes, workplaces, cities, and daily routines.
- Both the challenges and opportunities that emerge from this technology adoption have also been explored. By offering a detailed overview of the emerging landscapes of ML and IoT, our goal is to equip the readers with a deeper understanding of these technologies transformative power and encourage continuous investigation into this vibrant area.

The structure of this study is as follows: In Section 9.1, the basic concepts of IoT and ML and how integrating ML with IoT helps to improve the predictability and performance of IoT devices are presented. Section 9.2 covers the applications of ML and IoT, followed by Section 9.3, which is about challenges and future directions, and Section 9.4 concludes the chapter.

9.2 ML- AND IoT-BASED APPLICATIONS

The convergence of IoT and ML technology revolutionizes many aspects of our everyday life significantly enhancing, productivity, efficiency and convenience with creative solutions. By merging intelligent data analytics with connected devices, ML-driven IoT applications transform the way we interact with everything around us, from the innovation in smart homes to the transformation of urban environments and beyond. These applications utilize the power of IoT and ML to make our environments more responsive, making life simpler and more connected. These applications use cutting-edge algorithms to glean insightful information from the massive volumes of data produced by IoT sensors, facilitating proactive real-time automation and decision-making [13]. Consequently, ML and the IoT have become an inseparable part of our daily lives, offering tailored experiences, maximizing the use of available resources, and clearing the path for a more intelligent and connected future, as shown in Figure 9.4. Next, we discussed several daily applications that rely on this integration.

Machines smart enough to perform tasks on their own are being developed using artificial intelligence. Large amounts of data are produced by IoT devices, which are used to create advanced AI models. The quality of human existence can be greatly improved by combining AI and IoT. Intelligent processes that use examples or past experiences to optimize

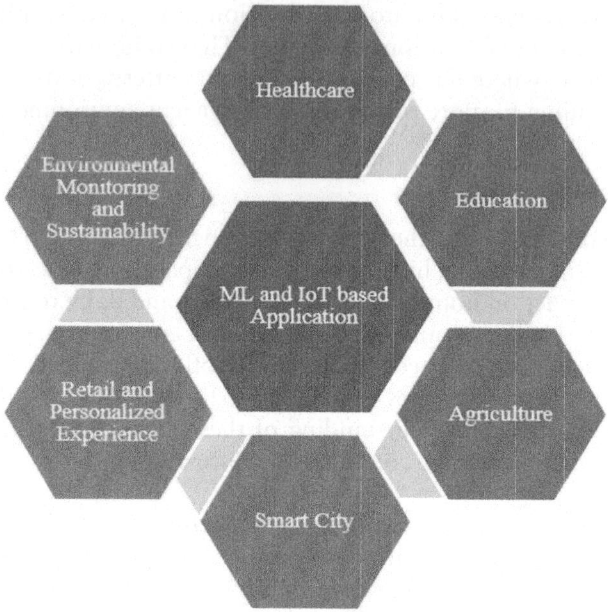

Figure 9.4 Applications based on ML and IoT collaboration.

performance criteria are known as ML. More specifically, ML algorithms use mathematical methods on massive datasets to create behavior models. Future estimates based on the recently obtained data are derived from these models.

9.2.1 Healthcare and wellness

ML and IoT technologies are revolutionizing patient care, wellness management, and medical research in the healthcare industry [14]. Wearable technology allows for continuous health tracking and early diagnosis of health issues. These gadgets are fitted with sensors that monitor vital signs, activity levels, and sleep patterns. This data is analyzed by ML algorithms, which then offer individualized insights, enable remote patient monitoring, and aid in clinical decision-making. Additionally, telemedicine platforms and IoT-enabled medical devices are expanding access to healthcare services, especially in underserved or remote places. Healthcare providers may provide more proactive, effective, and patient-centered treatment by utilizing the potential of ML and IoT [15].

9.2.1.1 Internet of medical things

Internet of medical things (IoMT) is a network of interconnected medical applications and devices that link to networks of healthcare information technology. This would lessen the demand for healthcare facilities and the number of needless hospital trips by allowing patients and doctors to communicate over a secure network. Wearables, medical/vital monitors, real-time location locators, telehealth, and other services fall under the umbrella of IoMT. These smart devices are intended for usage exclusively on the body and can be used at home, in clinics, or community settings [16].

9.2.2 Smart home and lifestyle

With the introduction of smart home technologies, our living spaces are now more connected, efficient, and convenient than ever. Smart homes can reduce energy consumption, improve security, and automate a variety of chores by utilizing ML algorithms and IoT devices. ML and IoT-based applications transform the way we interact with our homes, introducing intelligent lighting systems that adapt their illumination based on the patterns of occupancy and smart thermostats designed to learn and adjust to user preferences. These advances and technology allow our living environments to become more responsive, energy efficient, and tailored to your individual needs. By analyzing data on how we live, these systems promise not only to enhance our comfort but also optimize energy use contributing to a more sustainable future [4]. Furthermore, consumers can now more easily operate their smart gadgets and personalize their living spaces to their tastes thanks to developments in voice assistants and home automation platforms.

9.2.2.1 Internet of Everything (IoE)

In order to make networked interactions more significant and meaningful than before, the Internet of Everything (IoE) "is putting people, systems, data, and items together; turning information into actions that generate new capacities, richer experiences, and unparalleled economic opportunities for companies, individuals, and countries."

9.2.3 Workplace productivity and efficiency

IoT and ML technologies are revolutionizing the way businesses operate, communicate, and innovate in the modern workplace. IoT-enabled smart office systems with ML algorithms and sensors maximize workspace usage, boost worker productivity, and increase workplace safety. IoT data is used by predictive maintenance systems to track the condition of equipment, identify abnormalities, and cut down on expensive downtime. Moreover, ML-driven analytics solutions enable businesses to extract valuable insights from massive data sets, facilitating data-driven strategic planning and decision-making [10]. ML and IoT will become more crucial as workplaces change in terms of promoting creativity, productivity, and worker well-being.

9.2.4 Agriculture

Among the wide range of IoT applications, IoT-based smart agriculture has piqued the interest of numerous researchers, who have utilized IoT and ML technology to conduct ground-breaking research [17]. Agricultural yields can be increased with the use of IoT-based data-driven farming management techniques by planning production costs, reducing losses, and making better use of resources [18]. The AI techniques has revolutionized farming by collecting data from all throughout an agricultural farm using IoT [19]. This data includes quality levels for photos and actuator states for robot positioning. Without high-quality data, ML algorithms cannot be used to create precise predictive models. Applying these algorithms to these upgraded datasets enables more accurate predictions and better analysis. Systems for IoT-based agriculture have been applied in several research articles. Each sensor mode in an IOT system typically includes a microprocessor, various sensors (from simple temperature sensors to cameras), actuators, and wireless interfaces. Large amounts of sensor data may be efficiently stored and handled using the IoT, which can also include cloud computing services like cloud storage and agricultural maps. Real-time data access from any location at any time enables real-time tracking and end-to-end connectivity across all devices.

9.2.5 Education

Education is crucial for human development, and developing countries are conducting research projects to improve their education strategies and policies. To improve individual learning resources, researchers are focusing on student assessment and integrating IoT sensors into classrooms to collect real-time data on student behavior, attendance and performance. This data can be combined with ML techniques to create personalized learning pathways, identify at-risk students and provide tailored interventions. IoT-powered smart classrooms include interactive displays, smartboards and connected devices that can assist in capturing learning experiences. Content is modified by ML algorithms and provides rapid feedback. IoT devices enable seamless remote learning and provide access to teaching materials outside the traditional classroom [6]. In this digital era IoT with ML aim to create a more effective, inclusive and flexible learning environment for both teachers and learners

9.2.6 Urban living and transportation

Cities across the globe are implementing smart city initiatives to address urban challenges as well as enhance citizens' living standards. The transformation in public services, transportation networks and urban infrastructure mainly relies on ML plus IoT technologies [2]. Smart traffic management systems use real-time data from sensors connected via the IoT to improve traffic flow, alleviate congestion and enhance road safety. Cities also use predictive analytics along with IoT-enabled sensors for energy monitoring, environmental risk surveillance as well as air pollution tracking. The employment of AI in urban infrastructure planning could help make society more sustainable, resilient, and a better place to live [4].

9.2.7 Retail and personalized experiences

Retailers are personalizing shopping, growing customer engagement and transforming operations through the use of IoT and ML technologies [3]. Using a variety of touch points like online interactions, in-store behavior and purchase history, retailers can adapt product recommendations, promotions and marketing campaigns to individual customers' preferences by analyzing their data. By applying IoT devices such as beacons and RFID tags this helps retailers monitor stock levels optimize store layout and minimize stockouts. Also merchant's world has been enhanced with technology that enables them to compete better in an online platform through demand forecasting, supply chain management, fraud detection using ML algorithms.

9.2.8 Environmental monitoring and sustainability

Deploying IoT sensors across a range of ecosystems including air, water and soil, makes it possible for real-time data on environmental issues such as pollution levels, temperature, humidity and biodiversity to be continuously transmitted and collected. Such ML algorithms can start from this huge amount of data to analyze and derive meaningful insights out of it. By examining patterns, trends in environmental data, the ML systems could undertake predictive modeling as well as early detection of environmental hazards [5]. Additionally, ML-driven algorithms have the potential to improve resource management strategies like conservation of water, waste management and energy consumption to support sustainable practices. Smart grids as well as intelligent transport systems are two examples of smart infrastructure enabled by IoT applications which can help to reduce power consumption along with carbon emission. This information can be analyzed using ML techniques aimed at improving energy distribution, demand forecasting and minimizing environmental impact. Moreover; through crowdsourced data collection and analysis actions, IoT together with ML might enable citizens to interact and participate in environment preservation initiatives. We can say that the use of IoT and ML in environmental monitoring and sustainability has proved revolutionary, as it has the potential to solve various environmental-related issues and thus provide a way to develop a sustainable future for forthcoming generations.

9.3 CHALLENGES AND FUTURE DIRECTIONS

Undoubtedly, ML and IoT have revolutionized various aspects of our lives, yet this widespread adoption also has various challenges that need to be addressed to embrace their deployment in an effective and responsible manner.

Data Security and Privacy: As all the data collection process is done through the IoT platform, exposing data to threats like side-channel and cross-site scripting. One of the primary concerns about IoT devices is the possibility of data leaks [20]. For example, in 2019, Capital One faced a huge data breach that exposed the personal information of over 100 million customers due to a misconfigured firewall on an AWS server, demonstrating the dangers of IoT devices [21]. IoT technologies, such as wearable health trackers and medical implants, collect sensitive patient information. If sensitive patient data acquired by IoT devices enters into the wrong hands as a result of a breach, it might be manipulated or misused to propagate disinformation, even mal-information. Any breach of sensitive data could result in serious privacy violations and potentially threaten lives [22].

Data Quality: IoT devices collect data by means of various sensors or sources, which may introduce inconsistencies, resulting in data inaccuracies. In today's tech era, data acts as fuel. As we live in a data-oriented era, data is now a new power. Furthermore, data from IoT devices needs to be integrated with existing data to derive meaningful insights. However, integrating data from several sources can be difficult due to variances in format, standards, and quality. Poorly integrated data might result in errors and inconsistencies in analysis and decision-making, as ML algorithms rely on high-quality and balanced data to generate reliable results, and skewed data generates distorted results [23].

Limited Dataset Availability: Limited availability of relevant and high-quality data that can be used to train ML models or optimize IoT systems, bounded the efficacy of these techniques. When ML algorithms are connected with IoT devices, they frequently use data obtained from sensors, devices, or other sources to create predictions, optimize operations, or provide insights. However, a lack of relevant datasets might make it difficult to effectively train ML models or derive significant insights from IoT data. The ML algorithm requires historical data on a specific domain or task in order to complete the specified task, whether it is identifying patterns or performing classification. However, a lack of public datasets providing significant data can impede the creation and effectiveness of the required system [23].

To address these challenges, it is required that concerned efforts be made by stakeholders across industries, academia, and government. As technology is emerging day by day and these trends in ML and IoT-based applications focus on various aspects of day-to-day life, Shift toward edge computing and federated learning to enable faster processing of data at the network edge while preserving privacy [22]. ML models are black box models, where we just know the decision but have no idea why that particular decision is made up, which led to explainable AI (XAI) techniques to provide transparency and accountability in decision-making processes [16]. Intelligent interfaces and augmented reality systems will improve efficiency and help, leading to a stronger focus on human-machine collaboration. The success of an application is determined by the level of trust the user has in it. The ease of use and security provided by the given app boost esteem among users and service providers. As the concepts of centralized and decentralized services take part in the integration of IoT devices and ML, it is required to have innovative, sustainable solutions from academia and industry researchers to raise concerns in ML and IoT-based cloud environments. As the ideas of centralized and decentralized services play a major role in the integration of IoT devices and ML, academic and industrial researchers must provide new, long-term solutions to address challenges in ML and IoT cloud environments. With cutting-edge security frameworks and preventative actions against cyber threats, protecting IoT ecosystems will continue to be an important objective. The incorporation of ML and IoT technologies will optimize resource use and promote eco-friendly behaviors across multiple sectors, culminating in a focus on sustainability.

9.4 CONCLUSION

The rapid growth of computer and communication technologies has made the IoT and ML increasingly pervasive in our daily lives. The cloud and the IoT are two evolved concepts that are designed to work together. IoT collects real-time data and stores it in the cloud, which users or consumers can access remotely at any time. This collected data is then used by artificial intelligence technologies, specifically Ml and DL, for decision-making or predictions. This chapter examined the practical applications of these approaches. However, these devices, gadgets, or solutions have several disadvantages, including security, privacy, trust, standardization, and so on. We also discuss these difficulties, as well as recommendations for how to address them. By embracing emerging trends and working toward a more interconnected, intelligent, and sustainable future, we may successfully leverage the revolutionary power of ML and IoT innovations to improve living standards to meet societal concerns.

REFERENCES

[1] Lee J, Stanley M, Spanias A, Tepedelenlioglu C. Integrating machine learning in embedded sensor systems for Internet-of-Things applications. In *2016 IEEE international symposium on signal processing and information technology (ISSPIT)* 2016 Dec 12 (pp. 290–294). IEEE.

[2] Mishra S, Tyagi AK. The role of machine learning techniques in internet of things-based cloud applications. *Artificial intelligence-based internet of things systems*. 2022: IEEE, 105–135.
[3] Al-Turjman F, Baali I. Machine learning for wearable IoT-based applications: A survey. *Transactions on Emerging Telecommunications Technologies*. 2022 Aug;33(8):e3635.
[4] Whaiduzzaman M, Barros A, Chanda M, Barman S, Sultana T, Rahman MS, Roy S, Fidge C. A review of emerging technologies for IoT-based smart cities. *Sensors*. 2022 Nov 28;22(23):9271.
[5] Bharadwaj HK, Agarwal A, Chamola V, Lakkaniga NR, Hassija V, Guizani M, Sikdar B. A review on the role of machine learning in enabling IoT based healthcare applications. *IEEE Access*. 2021 Feb 16;9:38859–38890.
[6] Laghari AA, Wu K, Laghari RA, Ali M, Khan AA. A review and state of art of Internet of Things (IoT). *Archives of Computational Methods in Engineering*. 2021 Jul:1–9.
[7] Gubbi J, Buyya R, Marusic S, Palaniswami M. Internet of Things (IoT): A vision, architectural elements, and future directions. *Future Generation Computer Systems*. 2013 Sep 1;29(7):1645–1660.
[8] Cao K, Liu Y, Meng G, Sun Q. An overview on edge computing research. *IEEE access*. 2020 May 1;8:85714–85728.
[9] Dhiman P, Kaur A, Bonkra A. Fake information detection using deep learning methods: a survey. In *2023 International conference on artificial intelligence and smart communication (AISC)* 2023 Jan 27 (pp. 858–863). IEEE.
[10] Sharifani K, Amini M. Machine learning and deep learning: A review of methods and applications. *World Information Technology and Engineering Journal*. 2023;10(07):3897–3904.
[11] Dhiman P, Kaur A, Iwendi C, Mohan SK. A scientometric analysis of deep learning approaches for detecting fake news. *Electronics*. 2023 Feb 14;12(4):948.
[12] Dhiman P, Bonkra A, Kaur A, Gulzar Y, Hamid Y, Mir MS, Soomro AB, Elwasila O. Healthcare trust evolution with explainable artificial intelligence: Bibliometric analysis. *Information*. 2023 Oct 3;14(10):541.
[13] Bharti S, Singh SN. Analytical study of heart disease prediction comparing with different algorithms. In *International Conference on Computing, Communication & Automation* 2015 May 15 (pp. 78–82). IEEE.
[14] Bhatia H, Panda SN, Nagpal D. Internet of things and its applications in healthcare-A survey. In *2020 8th International Conference on Reliability, Infocom Technologies and Optimization (Trends and Future Directions)(ICRITO)* 2020 Jun 4 (pp. 305–310). IEEE.
[15] Dhiman P, Bonkra A, Sandhu JK, Kaur A, Kaushal C. Analyzing breast cancer detection: A Scopus-driven study on the role of machine learning and deep learning. In *2023 6th International Conference on Contemporary Computing and Informatics (IC3I)* 2023 Sep 14 (Vol. 6, pp. 890–895). IEEE.
[16] Dhiman P, Bonkra A, Kaur A, Gulzar Y, Hamid Y, Mir MS, Soomro AB, Elwasila O. Healthcare trust evolution with explainable artificial intelligence: Bibliometric analysis. *Information*. 2023 Oct 3;14(10):541.
[17] Bonkra A, Pathak S, Kaur A, Shah MA. Exploring the trend of recognizing apple leaf disease detection through machine learning: A comprehensive analysis using bibliometric techniques. *Artificial Intelligence Review*. 2024 Feb;57(2):1–26.

[18] Kaur P, Singh MP, Mishra AM, Shankar A, Singh P, Diwakar M, Nayak SR. DELM: Deep ensemble learning model for multiclass classification of super-resolution leaf disease images. *Turkish Journal of Agriculture and Forestry.* 2023;47(5):727–745.
[19] Kaur P, Mishra AM, Goyal N, Gupta SK, Shankar A, Viriyasitavat W. A novel hybrid CNN methodology for automated leaf disease detection and classification. *Expert Systems.* 2024; 23, Feb 11:e13543.
[20] Bonkra A, Dhiman P. IoT security challenges in cloud environment. In *2021 2nd International Conference On Computational Methods in Science & Technology (ICCMST)* 2021 Dec 17 (pp. 30–34). IEEE.
[21] Novaes Neto N, Madnick S, de Paula MG, Malara Borges N. A case study of the capital one data breach. In Stuart E. and Moraes G., De Paula, Anchises and Malara Borges, Natasha, *A case study of the capital one data breach.* 2020 Jan 1, IEEE.
[22] Sharma B, Hooda S. AI-based prediction models for network security attacks: A comparative study. Available at SSRN 4638515. 2023 Nov 20.
[23] Dhiman P, Kaur A, Hamid Y, Ababneh N. Fake news detection datasets: A review and research opportunities. *International Journal of Computing and Digital Systems.* 2023 Jul 16;14(1):1–20.

Chapter 10

Skill verification system using blockchain

Architecture, challenges and proposed model

Sarah, Arun Kumar Rana, Roshni Jeewani and Pooja Gupta

10.1 INTRODUCTION

In the era of rapid digitalization and the gig economy, accurately assessing and verifying an individual's skills and competencies has become an essential aspect of modern professional and educational landscapes. Traditional methods of skill verification often suffer from issues of trust, security, and transparency. Blockchain technology is a ground-breaking breakthrough that has the potential to transform the way we authenticate and document skills [1].

Introducing a skill verification system powered by blockchain signifies a significant advancement in how we validate credentials. In response to the challenges posed by traditional verification methods, this innovative system harnesses the capabilities of blockchain technology to establish a decentralized, transparent, and secure framework for authenticating skills. The concept of immutability, inherent in blockchain, ensures that once skills are verified and documented, the information becomes resistant to tampering or fraudulent activities [2–5].

Operating on a decentralized network, the system eliminates the dependency on a central authority, encouraging direct peer-to-peer verification that is both efficient and trustworthy. Transparency is a foundational principle, with every skill verification transaction meticulously recorded on a publicly accessible ledger. Automation is facilitated through the use of smart contracts, which enable the seamless execution of verification processes when skills are endorsed or certified.

Privacy is a key consideration, and cryptographic techniques are employed to ensure secure data handling. Individuals retain control over who has access to their skill credentials, striking a balance between transparency and personal privacy. Moreover, the system emphasizes interoperability, allowing for smooth integration with existing systems.

In essence, this blockchain-powered Skill Verification System stands as a promising development for a future where skills are validated with unparalleled credibility. Its benefits extend to individuals, employers, and educational

 DOI: 10.1201/9781003590439-10

institutions, promising a more trustworthy and efficient means of assessing and confirming professional capabilities [6–11].

10.2 TRADITIONAL SKILL VERIFICATION SYSTEM

In the traditional system of issuing skill certificates, such as those provided by universities, the journey from completion of coursework to actual verification can feel like a labyrinthine odyssey. Picture this: a student diligently completes their studies, undergoes evaluations, and finally receives a paper certificate as proof of their hard-earned skills. Yet, when it comes time to verify these credentials, the process becomes a cumbersome ordeal as we can see in Figure 10.1. Potential employers or other institutions must navigate through a maze of contacting the issuing institution, waiting for responses, and hoping for accurate verification—all of which can be time-consuming and frustrating. Trust, too, hangs delicately in the balance, as parties must rely on the integrity of often opaque verification procedures. Contrast this with a skill verification system powered by blockchain technology, where the narrative takes on a distinctly different tone. Here, the story begins with the instantaneous recording of a student's achievements upon completion of their course or certification. This record, stored on an immutable blockchain ledger, serves as an indelible testament to their skills and capabilities. No longer must individuals wait anxiously for verification; with a simple query to the blockchain, their qualifications are readily accessible, eliminating the need for intermediaries or protracted verification processes. Trust, once a precarious commodity, becomes an inherent feature of this system, as the decentralized nature of blockchain ensures transparency and reliability. In this blockchain-enabled paradigm, time and resources are saved, as the verification process is streamlined to a fraction of its former self. The globe shrinks, figuratively speaking, as individuals can effortlessly validate their qualifications across borders and oceans. What once felt like an arduous journey through a labyrinth now resembles a swift and seamless passage—a testament to the transformative power of blockchain in revolutionizing traditional systems of skill verification [12–19].

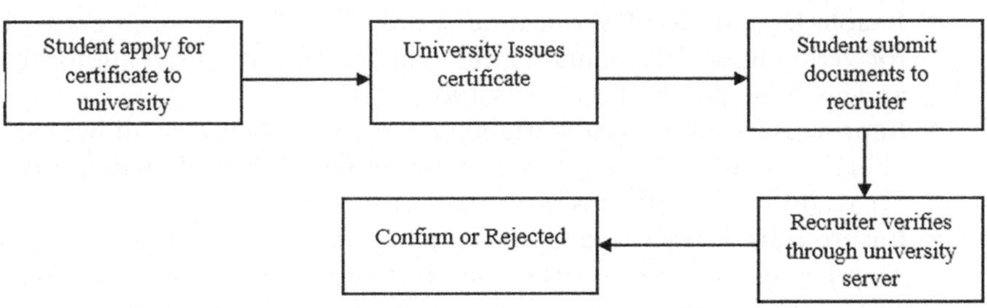

Figure 10.1 Depicts the flow of traditional system which is tedious and time-consuming.

10.3 PROBLEM STATEMENT

The Current skill verification process suffers from several shortcomings, including:

- **Lack of Trust**: Employers and institutions often face difficulties in verifying the authenticity of candidates' claimed skills and qualifications due to reliance on centralized databases or paper-based credentials. This lack of trust can lead to hiring inefficiencies and mismatches.
- **Vulnerability to Fraud**: Traditional verification systems are susceptible to fraud, as credentials can be easily forged or manipulated. This poses significant risks for employers, educational institutions, and other stakeholders relying on verified skills and qualifications.
- **Inefficiency and Delays**: The manual nature of skill verification processes results in inefficiencies and delays, consuming valuable time and resources. This hampers the agility of recruitment processes and academic admissions, impacting overall productivity and competitiveness.
- **Limited Accessibility**: Access to centralized verification systems may be restricted or unavailable in certain regions, particularly in developing countries or rural areas. This limits opportunities for individuals to showcase their skills and qualifications on a global scale.

10.4 OBJECTIVES

- **Enhance Trust and Transparency**: Implement a blockchain-based skill verification system to increase trust and transparency in the validation of credentials and qualifications.
- **Reduce Fraud and Manipulation**: Utilize blockchain's immutability and cryptographic security features to mitigate the risk of fraud and manipulation in the verification process.
- **Streamline Verification Processes**: Develop efficient and streamlined processes for verifying skills and qualifications, reducing administrative overhead, and improving the overall user experience.
- **Enable Decentralized Verification**: Establish a decentralized network for verifying skills, eliminating the reliance on centralized authorities and enabling peer-to-peer verification.
- **Improve Accessibility and Portability**: Create a system that allows individuals to easily access and share their verified skills and qualifications across different platforms and organizations.
- **Ensure Data Privacy and Confidentiality**: Implement robust data privacy measures to protect the confidentiality of individuals' personal and sensitive information while ensuring compliance with relevant regulations such as GDPR.

- **Facilitate Interoperability**: Ensure that the skill verification system is compatible with established credentialing systems and standards, facilitating effortless integration with diverse platforms and applications.
- **Enable Real-time Verification**: Enable real-time verification of skills and qualifications, allowing employers, educational institutions, and other stakeholders to instantly validate credentials as needed.
- **Empower Individuals**: Empower individuals to take ownership of their skills and qualifications by providing them with tools and mechanisms to manage and share their verified credentials securely.
- **Promote Lifelong Learning**: Encourage continuous learning and skill development by providing incentives and rewards for individuals to undergo skill verification and validation processes regularly.

10.5 LITERATURE REVIEW

The project aims to provide a Skill verification system. We referenced a few previously published papers and publications. Various individuals work in this subject. Our Literature Survey focused on Blockchain Technology.

1. Priti P. Bokariya and Dilip Motwani [2] designed a system on a blockchain-based credential verification system that enables digital certificate verification and validation. The authorities will scan a certified document with a QR code and a unique number issued by the user to validate the document, making the procedure faster and more efficient than the manual verification method. As a result, users do not need to bring documents with them; instead, they only need to submit the QR code and unique number found on certified documents.
2. Jerinas Gresch, Bruno Rodrigues, Eder Scheid, Salil S. Kanhere and Burkhard Stiller [5] introduced blockchain-based diploma management system named UZHBC (University of ZuricH BlockChain). It takes a diploma certificate as an input in the UZHBC frontend and generates a hash function corresponding to this, and this hash will be compared to the hash stored in smart contracts. If this is present then it is verified; otherwise, the system will also provide feedback.
3. Kumar, Dinesh and D S, Manoj [3] created a Blockchain-based Educational Certificate Verification System in 2020. This system stores academic degrees and awards digitally, preventing the use of fraudulent credentials. The method will employ a double SHA-256 algorithm to create a hash that, if validated, will be put as a block on the blockchain. It will also use a census technique to accept and reject documents.
4. Muhammad Dhiyaul, Rakin Zainuddin and Kan Yeep Choo [9] developed a blockchain-based document verification system that integrates with IPFS (Interplanetary File System).IPFS is a handy technology for

storing huge files and ensuring their integrity. Document verification relies heavily on file integrity.

5. Jiin-Chiou Cheng, Narn-Yih Lee, Chien Chi, and Yi-Hua Chen [6] provide an in-depth review of research on how blockchain may be used to verify documents. In this, authorities award documents, and the system checks the data and sends e-certificates instead of paper copies with QR codes that may be used to prove the legitimacy of documents.
6. Chukowry Varshinee, Nanuck Geeaneswari, and Sungkur Roopesh [7] Blockchain technology powers a digital badge and microcredential system. This work develops and presents a system in which a digital badging and micro-credential system consisting of quizzes and digital badges is constructed using the Ethereum blockchain, IPFS to store the badge picture, and Firebase for examination, registration, and saving courses.
7. **Paul Sarda, Mohammad Jabed Morshed Chowdhury, Alan W. Colman, Ashad Kabir, Jun Han** [10] introduced blockchain for fraud prevention that seeks to enhance work history verification by blockchain. This technology will also enable potential employees to combine several documents and submit them to prospective employers, who will be able to validate each record against the blockchain version while retaining both sides' anonymity.
8. **Alaa D. Alrehily, Asmaa F. Alotaibi, Suzan B. Almutairy, Mashael S. Alqhtani, Jayaprakash Kar** [13] introduced the notion of digital signatures to assure security. Digital signatures may be created using techniques such as RSA, DSA, and ECDSA. It takes a detailed look at how digital signatures may offer authenticity and integrity. It is used in cryptocurrencies such as Bitcoin and Ethereum. It is also utilized in RFID and the Internet of Things.
9. **J. Gupta and S. Nath** [18] introduced the Incentive-based Certification system using blockchain named as SkillCheck. SkillCheck is a fresh idea for online education that uses a special kind of technology called blockchain to solve problems like large class sizes and fair grading. It lets people take courses whenever they want, not just during traditional school times. SkillCheck tries to make sure everyone gets graded fairly by using a special system that stops biases from affecting grades.
10. **Gunit Malik, Kshitij Parasrampuria, Sai Prasanth Reddy, Seema Shah** [8] introduced a blockchain-based identity verification model for verifying documents with the use case of the government, such as pass ports, driving licenses, etc. It provides a platform for issuing agencies to check the validity of supplied documents via a rapid, reliable, and secure method. This would significantly speed up the issuance of new papers by eliminating the requirement to get documents from other issuing bodies and have them certified.

10.6 PROPOSED WORK

Our proposed work involves the development and implementation of a skill verification system using blockchain technology. This system aims to revolutionize the validation of skills and qualifications by leveraging the decentralized, transparent, and secure nature of blockchain.

Firstly, we will conduct thorough research and analysis to understand the current challenges and requirements of skill verification processes. Collaborating closely with stakeholders, including employers, educational institutions, and individuals, we will gather insights to inform the design and development of the system. Our approach will prioritize interoperability, ensuring seamless integration with existing credentialing systems and standards. This will involve careful planning for the compatibility of data structures, protocols, and interfaces to facilitate smooth interaction with diverse platforms and applications.

The system's architecture and framework will be meticulously designed to accommodate blockchain principles while addressing scalability, security, and usability considerations. Smart contracts will automate the verification process, ensuring efficiency and reliability in credential validation. During the development phase, we will focus on creating intuitive user interfaces and robust smart contracts, backed by rigorous testing and quality assurance procedures. Pilot implementation will allow us to validate the system's performance and gather feedback from real-world usage.

Upon deployment, we will closely monitor the system's operation, implementing regular maintenance activities and addressing any issues promptly. Continuous evaluation and feedback will drive iterative improvements, ensuring that the skill verification system remains effective and responsive to evolving needs.

The Figure 10.2 gives us the flow of different parties in the system how each party will work. Starting with the user- the user will upload his documents on system and then he would request the universities/institution to endorse it. Now with our system we can know that is a valid course that a person has done. Then the person uploading the documents can also ask to endorse it and ask other universities and institution to endorse it. With that, increases the percentage of his skills to get more acknowledged.

After endorsement, those skills will be displayed on the candidate's information that he has been endorsed by this this institution. Also, Organization would be able to add new employee if needed and they can edit the list according to their need [20–27].

Moving to Figure 10.3, it shows the basic flow-chart of our system. Starting with the login we then request to add a user to that blockchain. Then the admin can either accepts or rejects to add a specific user in the blockchain. The user will be added to the blockchain by entering all the information like name, organization he went to, certificates he has and with all those information the hash pf that information is calculated and then the

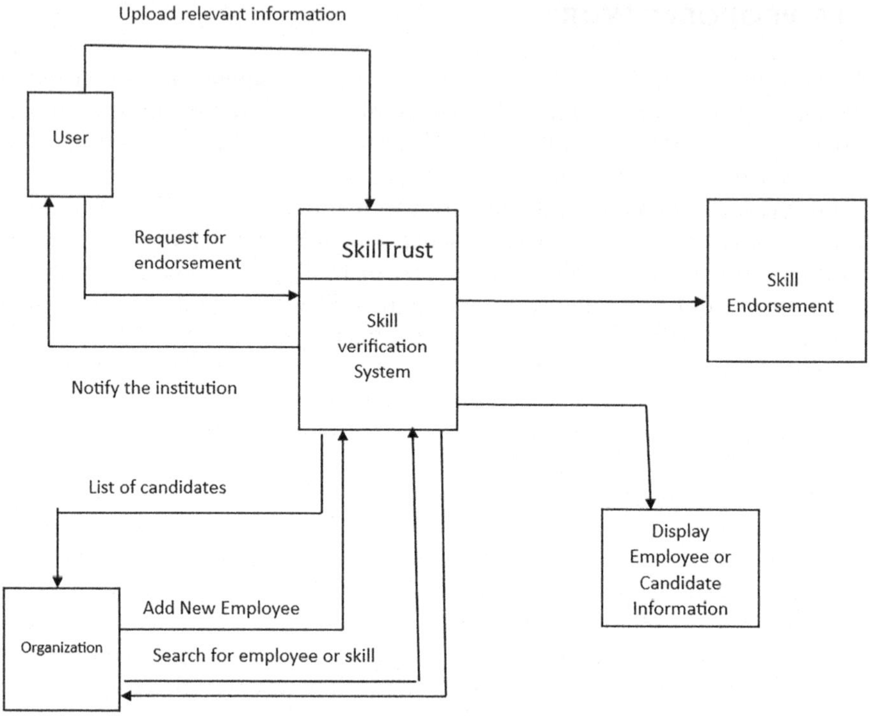

Figure 10.2 Flowchart of skill verification system.

hash is saved in the blockchain. The hash is a one-way function which can give an output in a single way only. There's no way to get the input back from the output. Moving forward then the user would be able to add that skill in their account and then the user can request the organization for the endorsement and the decision will be in the organization's hand weather they want to endorse his skill or not. Skill is endorsed? Congratulations you just get more visibility than others [28–32].

10.7 TECHNOLOGY USED

10.7.1 Frontend

- **React.js:** React.js is a JavaScript framework and library developed by Facebook. When compared to simple JavaScript, it enables faster and more efficient creation of interactive user interfaces and web applications. React is only concerned with state management and displaying information to the DOM, therefore creating React apps typically needs the usage of other frameworks for routing and particular client-side functionality.

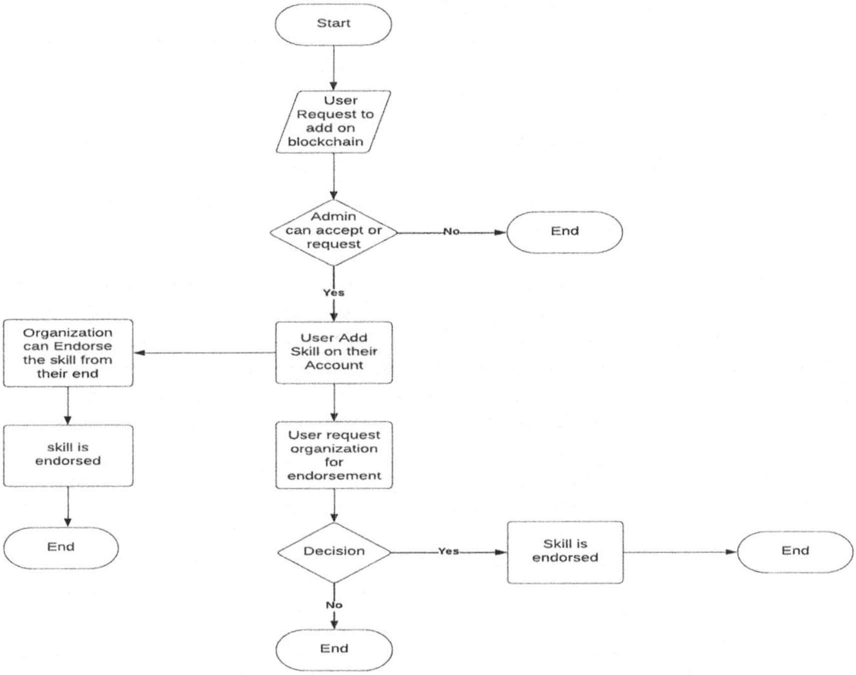

Figure 10.3 Flowchart of skill verification system.

- **Metamask**: MetaMask must be installed before it can be used as an Ethereum wallet, exactly as other browser plugins. Once loaded, users may conduct transactions by saving Ethereum and other ERC-20 tokens to any Ethereum address. Dapps instantly connect to MetaMask, shortening the connection time. If payment is necessary within the DApp, a pop-up window will appear prompting you to approve the transaction from your MetaMask account.

10.7.2 Backend

- **Ethereum**: Ethereum, founded in 2015, is a blockchain-based, open-source, decentralized computing platform with its own currency, ether. It allows you to construct and administer Distributed Applications and Smart Contracts without interference from third parties, fraud, or control. Ethereum is a blockchain-based platform and a programming language (Turing complete) that allows for the development and dissemination of distributed applications.
- **Smart Contracts**: Smart contracts in blockchain programming use basic "if/when…then" statements. When certain conditions are met, a network of computers will carry out the operations. They include distributing notices, collecting money, registering vehicles, and issuing

Table 10.1 Depicts challenges of skill verification system

Challenges	*Description*
Scalability	Blockchain networks can face scalability challenges, especially during periods of high demand. As the number of participants and transactions increases, some blockchain networks may experience slower transaction processing times and higher fees.
Data Privacy	Storing personal and sensitive information on a blockchain may raise privacy concerns. While blockchain provides immutability and security, it's essential to ensure that private information is adequately protected, and only authorized parties have access.
Reliability on Smart Contracts	Bugs or vulnerabilities in smart contracts could compromise the reliability of the skill verification process. Smart contracts, which automate processes on the blockchain, must be thoroughly tested and audited to avoid errors or security vulnerabilities.
Adoption & Standardization	Getting widespread adoption of blockchain-based verification systems can be difficult. Establishing universal standard for skill verification on the blockchain is crucial for interoperability and broader acceptance.

tickets. When the transaction is successful, the blockchain is updated. As a result, the transaction is irreversible, and only those with authority may view the results. Simply said, smart contracts are algorithms that run on blockchains when specific conditions are met.

- **Ganache**: "Ganache" is a software tool used mostly in Ethereum development. Ganache offers a personal blockchain that allows you to run tests, issue commands, and monitor state while still having control over the blockchain's functioning. It is frequently used to run smart contracts and decentralized apps in a local development environment.

10.8 CHALLENGES AND LIMITATIONS

Implementing a skill verification system using blockchain technology offers several advantages, such as transparency, immutability, and decentralization. However, at the same time it comes with its own set of challenges and limitations [33–36]. Here are some of them are as follows in Table 10.1.

10.9 CONCLUSION

The system proposes a decentralized, transparent, and reliable framework for online Skill Tests using Blockchain technology The time spent on competency checks and skill verification may be decreased with the use of a blockchain-based system, which can also increase internal organization trust in the management of skills and competencies. This system can be expanded to ensure integrity in government, education, and private sectors. The long-term goal is to test and evaluate the proposed framework in a real-world setting.

REFERENCES

[1] Harshita Bhosale, Rutuja Kanki and Gayatri Jaiswal, "Revolutionizing Verification and Management of Educational Certificates with Self-Sovereign Student Identities using Blockchain", *International Research Journal of Engineering and Technology (IRJET)*, vol. 08, no. 5, May 2021.

[2] Priti P. Bokariya and Dilip Motwani, "Decentralization of Credential Verification System Using Blockchain", *International Research Journal of Engineering and Technology (IRJET)*, vol. 10, no. 11, September 2021.

[3] K Dinesh Kumar, P Senthil, D S Manoj Kumar, "Educational Certificate Verification System Using Blockchain", *International Journal of Scientific & Technology Research*, 2, 11–23 January 2023.

[4] Rowthu Ramya, Satyam Kasturi Lakshmi, Atkuri Satya Venkata Surya Naga Sri Sai, Indana Yaswanth V and Appa Rao, "Blockchain Based Academic Certificate Authentication", *Journal of Engineering Sciences*, vol. 13, no. 5, May 2022.

[5] Jerinas Gresch, Bruno Rodrigues, Eder Scheid, Salil S. Kanhere and Burkhard Stiller, "The Proposal of a Blockchain-based Architecture for Transparent Certificate Handling", *1st Workshop on Blockchain and Smart Contract Technologies*, July 2018.

[6] Jiin-Chiou Cheng, Narn-Yih Lee, Chien Chi and Yi-Hua Chen, "Blockchain and Smart Contract for Digital Certificate", *2018 IEEE International Conference on Applied System Innovation (ICASI)*, April 2018.

[7] Varshinee Chukowry, Geeaneswari Nanuck and Roopesh Kevin Sungkur, "The Future of Continuous Learning-Digital Badge and Microcredential System Using Blockchain", *Global Transitions Proceedings*, vol. 2, November 2021.

[8] Gunit Malik, Kshitij Parasrampuria, Sai Prasanth Reddy and Seema Shah, "Blockchain Based Identity Verification Mode", *2019 International Conference on Vision Towards Emerging Trends in Communication and Networking (ViTECoN)*, March 2019.

[9] Muhammad Dhiyaul Rakin Zainuddin and K.Y. Choo, "Design a Document Verification System Based on Blockchain Technology", *Proceedings of the Multimedia University Engineering Conference*, December 2022.

[10] A. Raja, "Blockchain for Fraud Prevention: A Work-History Fraud Prevention System", *2018 17th IEEE International Conference On Trust, Security And Privacy In Computing And Communications/ 12th IEEE International Conference On Big Data Science And Engineering (TrustCom/BigDataSE)*, August 2018.

[11] Md. Shahriar Karim Shawon, Hosnain Ahammad, Shumrose Zaman Shetu, Md. Mahfujur Rahman and Syed Akhter Hossain, "A Blockchain-Based Solution for Verification of Educational Certificates", *12th International Conference on Computing Communication and Networking Technologies (ICCCNT)*, July 2021.

[12] T. Rama Reddy, P.V.G.D. Prasad Reddy, R. Srinivas et al., "Proposing a Reliable Method of Securing and Verifying the Credentials of Graduates Through Blockchain", *EURASIP J. on Info. Security*, vol. 7, June 2021.

[13] Alaa D. Alrehily, Asmaa F. Alotaibi, Suzan B. Almutairy, Mashael S. Alqhtani and Jayaprakash Kar, "Conventional and Improved Digital Signature Scheme: A Comparative Study", Vol. 1, *Journal of Information Security*, January 2015.

[14] Nero Chaniago, Parman Sukarno and Aulia Wardana, "Electronic Document Authenticity Verification of Diploma and Transcript Using Smart Contract on Ethereum Blockchain", *Register: Jurnal Ilmiah Teknologi Sistem Informasi*, vol. 7, pp. 149, May 2021.

[15] Liyuan Liu, Meng Han, Yiyun Zhou and Reza M. Parizi, "E^2 C-Chain: A Two-Stage Incentive Education Employment and Skill Certification Blockchain", *2019 IEEE International Conference on Blockchain (Blockchain)*, pp. 140–147, July 2019.

[16] Koteru Ramani Navya, Yash Patel and Parminder Kaur Makode, "Blockchain Application in Incentivizing Students on Participation in Online Classes", *2022 IEEE 2nd Mysore Sub Section International Conference (MysuruCon)*, pp. 1–7, November 2022.

[17] Mahran Morsidi, Sharul Tajuddin, Ravi Kumar Patchmuthu and S. H. Shah Newaz, "Blockchain-Based Reward System: A Means for Providing Incentive to Students for Teaching Feedback", *2021 International Conference on Electronics, Communications and Information Technology (ICECIT)*, pp. 1–5, November 2021.

[18] Jay Gupta and Swaprava Nath, "SkillCheck: An Incentive-based Certification System using Blockchains", *2020 IEEE International Conference on Blockchain and Cryptocurrency (ICBC)*, pp. 1–3, November 2020.

[19] Osman Ghazali and Omar S. Saleh, "A Graduation Certificate Verification Model via Utilization of the Blockchain Technology", *Journal of Telecommunication, Electronic and Computer Engineering*, vol. 10, no. 3–2, September 2018.

[20] Ravi Singh Lamkot, Devdoot Maji and Hitesh Shetty, "Certificate Verification using Blockchain and Generation of Transcript", *International Journal of Engineering Research & Technology (IJERT)*, vol. 10, no. 03, March 2021.

[21] Q. Aini, U. Rahardja, M. R. Tangkaw, N. P. L. Santoso and A. Khoirunisa, "Embedding a Blockchain Technology Pattern Into the QR Code for an Authentication Certificate", *Jurnal Online Informatika*, vol. 5, no. 2, December 2020.

[22] A. Rana, S. Sharma, K. Nisar, A. Ibrahim, A. Ag, S. Dhawan … and N. Goyal, "The Rise of Blockchain Internet of Things (BIoT): Secured, Device-to-Device Architecture and Simulation Scenarios", *Applied Sciences*, vol. 12, no. 15, p. 7694, 2022.

[23] A. Kumar, S. Sharma, N. Goyal, A. Singh, X. Cheng and P. Singh, "Secure and Energy-Efficient Smart Building Architecture With Emerging Technology IoT", *Computer Communications*, 176, 207–217, 2021.

[24] A. Kumar, S. Sharma, A. Singh, A. Alwadain, B. J. Choi, J. Manual-Brenosa, … and N. Goyal, "Revolutionary Strategies Analysis and Proposed System for Future Infrastructure in Internet of Things", *Sustainability*, vol. 14, no. 1, p. 71, 2021.

[25] A. Rana, C. Chakraborty, S. Sharma, S. Dhawan, S. K. Pani and I. Ashraf, "Internet of Medical Things-Based Secure and Energy-Efficient Framework for Health Care", *Big Data*, vol. 10, no. 1, pp. 18–33, 2022.

[26] A. Kumar and S. Sharma, "Internet of Robotic Things: Design and Develop the Quality of Service Framework for the Healthcare Sector Using CoAP", *IAES International Journal of Robotics and Automation*, vol. 10, no. 4, p. 289, 2021.
[27] S. Dhawan, C. Chakraborty, J. Frnda, R. Gupta, A. K. Rana and S. K. Pani, "SSII: Secured and High-Quality Steganography Using Intelligent Hybrid Optimization Algorithms for IoT", *IEEE Access*, vol. 9, pp. 87563–87578, 2021.
[28] S. K. Rana, S. K. Rana, A. K. Rana and S. M. Islam, "A Blockchain Supported Model for Secure Exchange of Land Ownership: An Innovative Approach", In *2022 International Conference on Computing, Communication, and Intelligent Systems (ICCCIS)* (pp. 484–489). IEEE, November 2022.
[29] A. K. Rana and S. Sharma, "Internet of Things Based Stable Increased-throughput Multi-hop Protocol for Link Efficiency (IoT-SIMPLE) For Health Monitoring Using Wireless Body Area Networks", *International Journal of Sensors Wireless Communications and Control*, vol. 11, no. 7, pp. 789–798, 2021.
[30] A. Kumar, S. Sharma, N. Goyal, S. K. Gupta, S. Kumari and S. Kumar, "Energy-Efficient Fog Computing in Internet of Things Based on Routing Protocol for Low-Power and Lossy Network With Contiki", *International Journal of Communication Systems*, vol. 35, no. 4, p. e5049, 2022.
[31] M. Pandit, D. Gupta, D. Anand, N. Goyal, H. M. Aljahdali, A. O. Mansilla … and A. Kumar, "Towards Design and Feasibility Analysis of DePaaS: AI Based Global Unified Software Defect Prediction Framework", *Applied Sciences*, vol. 12, no. 1, p. 493, 2022.
[32] U. K. Lilhore, A. L. Imoize, C. C. Lee, S. Simaiya, S. K. Pani, N. Goyal … and C. T. Li, "Enhanced Convolutional Neural Network Model for Cassava Leaf Disease Identification and Classification", *Mathematics*, vol. 10, no. 4, p. 580, 2022.
[33] S. K. Rana, A. K. Rana and S. Dhawan, "A Vital Fusion of Internet of Medical Things and Blockchain to Transform Data Privacy and Security", In *Convergence of Deep Learning and Artificial Intelligence in Internet of Things* (pp. 293–308). CRC Press, 2022.
[34] Rana, A. K., and Sharma, S., "Enhanced Energy-Efficient Heterogeneous Routing Protocols in WSNs for IoT Application", *IJEAT*, *9*(1), 4415–4418, 2019.
[35] Rana, A. K., and Sharma, S., "Industry 4.0 Manufacturing Based on IoT, Cloud Computing, and Big Data: Manufacturing Purpose Scenario", In *Advances in Communication and Computational Technology* (pp. 1109–1119). Springer, Singapore 2021.
[36] A. K. Rana, R. Krishna, S. Dhwan, S. Sharma and R. Gupta, "Review on Artificial Intelligence With Internet of Things-Problems, Challenges and Opportunities", In *2019 2nd International Conference on Power Energy, Environment and Intelligent Control (PEEIC)* (pp. 383–387). IEEE, October, 2019.

Chapter 11

Integration of blockchain technology for digital asset management

Sunny Kuhar, Sumit Kumar Rana, Manisha and Jaya

11.1 INTRODUCTION

In our conventional business, social, and political systems, various contracts, deals, and financial transactions are managed and documented in a predefined format. Because of the digitization and rapid proliferation of Internet technology, our world is quickly approaching one where end consumers demand transparency as a fundamental expectation. In the contemporary digital era, whether in business or other types of communication, all parties involved desire to transact without the need for an additional party and expect technological design to be trustworthy and reliable. It has been said that a blockchain will miraculously accomplish these goals. It is first utilized with cryptocurrencies like Bitcoin.

The development of blockchain technology is shown in Figure 11.1. From version 1.0 to version 4.0, the Blockchain's evolution has increased dramatically. The first two stages of the evolution are represented by Blockchain 2.0, where its environment can be created using smart contracts like Ethereum, and Blockchain 3.0, where the technology became applications centric and permeated daily life by enabling a variety of industries like medical, education, farming, e-commerce, and many more. Blockchain 1.0 was restricted to storing and transferring value (such as with Dash, Ripple, and Bitcoin). These enterprise blockchains include Ethereum Quorum, R3 Corda, and Hyperledger. Next, nearly all of the restrictions from the previous one are eliminated in Blockchain 4.0. Blockchain 4.0 makes use of a distributed environment that has significant problems with scalability and transaction rate limitations. It can manage latency, throughput, and scalability. A prime example of that is Rchain.

There are three different types of implementation environments that can use the blockchain [1].

1. **Permissioned Blockchain** [2]: Proprietary networks, also known as closed or private networks, are available in this environment and determine the participants' roles. Industries in particular have created this for own commercial use.

DOI: 10.1201/9781003590439-11

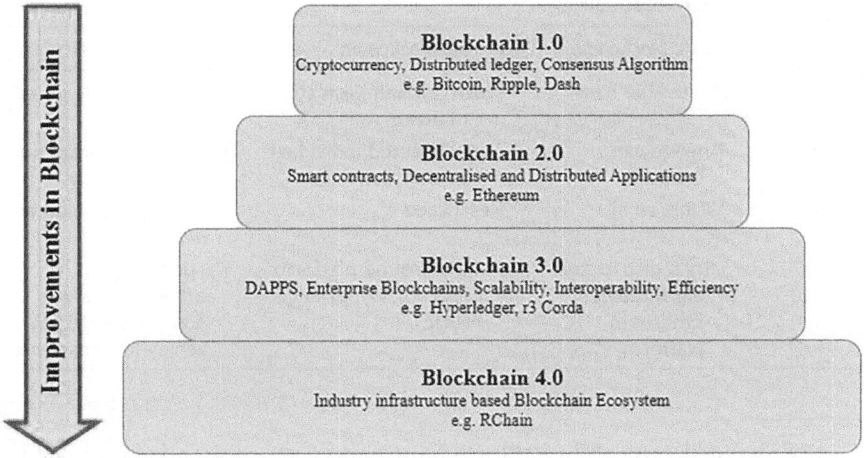

Figure 11.1 Evolution of Blockchain Technology.

2. **Permissionless or Public Blockchain:** Everyone is welcome to access, use, and engage in this open-source environment. E.g. Bitcoin Blockchain [3].
3. **Hybrid or Consortium Blockchain:** A third type of blockchain is referred as hybrid/consortium blockchain. It comes from fundamental Blockchain that were previously discussed. The number of participants determines the check over read and write access to data in a consortium blockchain. Groups of businesses or organizations that work together on projects use it. As a result, they work in an environment with limited access to complete their tasks and benefit from the consortium's use of technology.

Under different environment conditions, Blockchain technology offers several advantages as follow (Table 11.1):

- **Transparency**: Every participant can see the financial transactions that are kept on the Blockchain. The distributed record, which is the foundation of the Blockchain, is a distributed copy of a record that is updated by multiple users and can only be modified through a process of consensus, which requires the consent of all parties that are allowed to alter the file.
- **Enhanced security**: Blockchain provides more security than other record-keeping systems in a number of ways. Following the agreement of all parties who are allowed, transactions are added. The transaction is encoded and safely connected to the previous block once it has been approved by all parties. The blocks that contain the transactions are secured using secure hashing techniques that are appended to every block. Therefore, tempering a block is nearly impossible because it also necessitates changing other blocks of same chain.

Table 11.1 Types of blockchain

	Public blockchain	*Private blockchain*	*Consortium blockchain*
Access	Accessible to all	Network administrators can invite	Restricted to selected members
Transact	Anyone can transact	Only Selected individuals can transact	Designated consortium members only
View	Visible to all	Restricted	Restricted to selected members
Type	Large, distributed, e.g. bitcoin, Ethereum platform	Middle ground platform: Quorum, Hyperledger Fabric	Participating organizations are equally engaged in achieving consensus

- **Enhanced traceability**: Blockchain makes process and data tracking simple. All parties can see transactions, which makes any operation traceable. This technology makes it simple for businesses involved in the supply chain to track their products.
- **Fast and Efficient**: The paperwork in a conventional system is laborious, lengthy, and vulnerable to human error. Blockchain automation makes the process faster, more effective, and independent of outside interference.
- **Cost-effective**: Profit and cost-effectiveness are critical for any business. This technology eliminates the need for a middleman or other third party, making it more affordable.

11.2 OVERVIEW OF BLOCKCHAIN

A general outline of blockchain, includes its definition, applications, architecture and assisted technologies is provided in this section.

Definition: The "block" refers to the grouping of data, including transactions and other related data, and the term "chain" refers to the link among these blocks using a hash code.. This technology is now more secure thanks to these cryptographically connected blocks. In short, blockchain is a technology that quickly and easily gives each participating node access to verifiable and accessible data control over a distributed or decentralized environment. The nodes cannot be validated or verified by a single, centralized authority. Instead, a node must prove its identity by resolving a mathematical conundrum known as a proof of work in order to be allowed to join a network. A block can be introduced by a node that completes a proof of work.

Architecture: Blockchain technology allows multiple parties to communicate with each other without the need for a middleman. Transactions and communications are validated and verified by individual nodes known as miners. The approved transactions are stored in a structure called a block. The transactions that have already been committed are required in order for the current transaction to be carried out. This technology helps to prevent or limit double-spending in the digital currency system. A block further splits into two parts:

1. **Block header**: There are three parts that make up the block header. The hash from the prior block, which connects this block to the one before it, is the first element. The mining statistics that went into creating the block are included in the second section. The Markle tree root, the final part, is just the hash of the present block and serves as the foundation for confirming the accuracy of every single transaction that is part of the block. We utilize the hash code from the previous block to create the hash code for the current block. Therefore, it is practically impossible for an attacker to change all of the hash codes in the remaining chains if they attempt to alter the block contents. The Blockchain becomes tampered proof as a result. A timestamp is a recorded time, along with nonce and mining difficulty are among the mining statistics [4]. According to [5], a non-leaf node contains the cryptographic hash of its child nodes in the Merkle tree, and a leaf node comprises the hash chain of data blocks where transactions are hashed and secured.
2. **Transactions list**: The list of authorized transactions makes up the second part of the block. The size of the transaction and the block determine how many transactions are in a block. Asymmetric cryptography is used for transaction authorization and authentication. A transaction cannot be removed or changed once it has been added to the chain. A chain of blocks (called a blockchain) is formed when blocks are chained together and each block contains a hash of the one before it. If a block is legitimate and has proof of work—a computationally challenging hash produced by the mining process—it will be accepted into the chain. Since it uses a secure hashing algorithm (like SHA256) with secure hash pointers that point to the previous hash, it makes sure that any changes made to one block will require recalculating all subsequent blocks.
3. **Orphan block**: Using the transactions list that have not been added yet, miners attempt to mine blocks independently. A block broadcasts to every other node in the network for verifications after it has been mined by a miner. The block with the highest consensus among the numerous blocks in the network will be approved for inclusion. The network later discards the other block or blocks because they are deemed orphan blocks. Orphan blocks may contain transactions that have not yet been taken into consideration, but they also contain some transactions that have already been added to the valid block (Figures 11.2 and 11.3).

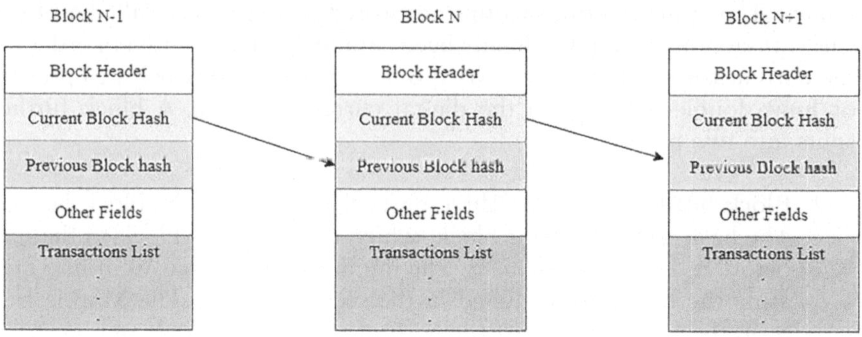

Figure 11.2 Blockchain architecture.

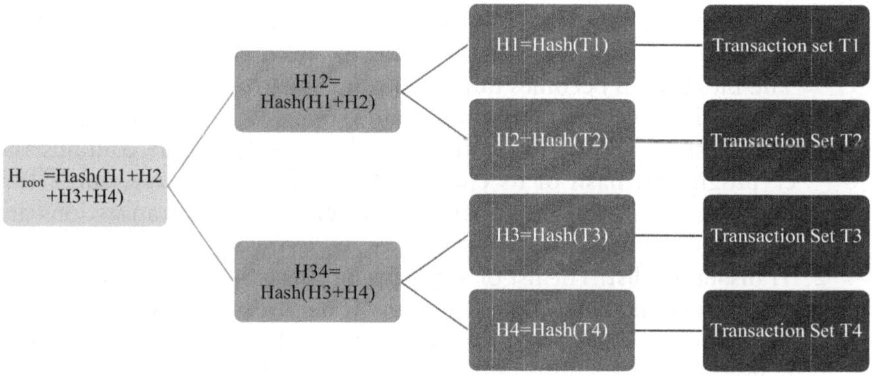

Figure 11.3 Merkle tree.

Fork: A fork is any chain that is not the legitimate one. An unmined block may occasionally be linked to the orphan chain, preventing it from joining the longest chain. A fork is created by these linked blocks.

Genesis block: The genesis block is the name given to the first block created in the system. The Genesis Block on the Bitcoin network is the first block that its creator, Satoshi Nakamoto, has ever mined.

11.3 DISTRIBUTED CONSENSUS IN BLOCKCHAIN

As exhibit in Figure 11.4, the algorithms are categorized into two classes.

Permissioned and Permission-less Blockchain: The majority of crypto-currencies that are currently on the market fall into the category of permission-less blockchains, which allow anyone to join the chain without the need for authentication or other restrictions. Without any censorship,

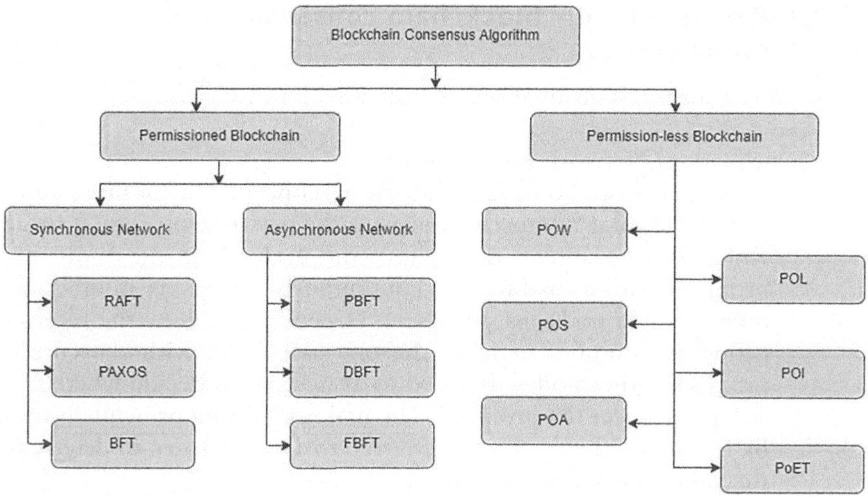

Figure 11.4 Consensus algorithm.

users can easily generate their own personal addresses and begin communicating with the Blockchain network. In a permissionless blockchain, anonymity, decentralization and transparency are the main concerns. [5]used Bitcoin as an example of a permission-less A few consensus mechanisms in the permission-less category are Blockchain and Proof of Work (PoW) [7], Proof of Stake (PoS) [8], Proof of Activity (PoA) [9], Proof-of-Location (PoL) [10], Proof-of-Importance (PoI) [11], and Proof-of-Elapsed-Time (PoET) [9].

On the other hand, a permissioned blockchain is a private network that users cannot join without authorization, censorship, or permission. In this Blockchain category, users are presumed to be acquainted with one another. Any organization, including private corporations and consortium groups, where certain authoritative entities are granted permission to participate, can use permissioned blockchain technology. A bug like a crash, network, or byzantine fault is the main obstacle to reaching distributed consensus in a permissioned blockchain. A permissioned blockchain example is Ripple [12].

Both synchronous and asynchronous networking scenarios are supported by this technology. The communication system in a synchronous environment needs to operate on a shared time slot with a finite delay, also referred to as a priori. Consensus mechanism utilized in permissioned Blockchain in synchronous environment are Paxis [13], Byzantine Fault Tolerance (BFT) [7], and RAFT [10]. Since there is no limit to delay in asynchronous environment like the Internet, time constraints have no place there. The latter is obviously more complex because it involves a variety of dynamic challenges for considerations like safety and liveliness, that will be covered in detail when we talk about each individual algorithm in the category.

11.3.1 Permissioned – blockchain consensus mechanism

These consensus mechanism are further divided into two categories.

i. **Synchronous Network Environment**
   - PAXOS: operates on a straightforward concept. One or more nodes out of the total number of nodes in the network propose a value, which is then spread throughout the network in the form of a proposal with a distinct and continuously increasing number. We refer to these nodes as proposers. Depending on how the received proposal's number matches the one associated with the current proposal, other nodes (referred to as acceptors) decide whether to accept or reject the proposal. The majority voting principle is used by the node's third category, referred to as the learner, to determine the value selected by acceptors.
   - RAFT With regard to performance and fault-tolerance as well as other aspects, RAFT was primarily created to function as an alternative to Paxos. Its primary concept is the splitting of major problems into smaller ones, which is handled separately. Together, all of the system's nodes choose a leader, and subsequent nodes follow the leader. The idea of majority vote is used to choose a leader from the pool of eligible candidates. The followers follow the leader's maintenance and replication of the state transition (such as logs). The leader never stops communicating its existence to all of its followers by putting out a unique message (called heartbeat). Because every other node follows the leader without question, the problem with RAFT is that the leader is assumed to be honest or correct.
   - BFT: Through a message-passing system, two army generals, one serving as commander and the other as a lieutenant—are in communication. Many scenarios, including one in which the commander is viewed as either faithful or a traitor, have been examined by classifying one or more lieutenants as either loyal or traitorous. In a system with three N nodes (generals), where a maximum of N nodes (generals) are faulty (traitor), consensus can be reached. This problem can be described as follows. A system that maintains tolerance against node failures associated with Byzantine faults is known as Byzantine Fault Tolerance.

ii. **Asynchronous Network Environment**
   - PBFT (Practical BFT): was first instigate by [14]. It was created for asynchronous communication environments similar to those found on the Internet, where there is no time constraint on when a request will be answered. Its goal was to resolve the problems brought up by the BFT process (e.g., not providing a result, providing results

that are intentionally false or misleading, etc.). The foundation of PBFT is state machine replication, with one node serving as the primary and others serving as secondary (slave/backup/follower) nodes that are chosen by a round-robin process.

- DBFT (Delegated BFT): According to [15], DBFT was created to address the performance and scalability issues that are the main causes for concern when implementing blockchain technology. Where N is the number of active nodes, the number of malfunctioning nodes in DBFT should not exceed [(N-1)/3]. Every group of small active nodes (consensus nodes) chooses a delegate by voting for their leader. While other nodes receive and verify blocks, all of these delegates strive to come to a consensus and produce fresh blocks. Among these delegates, there will be one overall leader who will make all of the decisions.
- FBFT (Federated Byzantine Fault Tolerance): The Blockchain platform's payment system makes use of this BFT variation. Stellar and Ripple [16] are two examples of such protocols. Since financial transactions are necessary for operation, any flaw or attack should be met head-on with FBFT. In FBFT, quorum slices are used to reach a consensus. When system-level quorums, or slices, form, the system is brought together by these slices. Open network participation is encouraged by FBFT, which results in natural network expansion.

11.3.2 Permissionless blockchain

- PoW (Proof of Work): According to [7] this is consensus process in which a computationally demanding mathematical problem is presented for solution. Examples of these are Bitcoin [5], Litecoin [4], Ethereum [4], etc. Its basic characteristic is that, while it may be challenging to solve, it is simple to verify (the right answer).
- PoS (Proof of Stake): [8] discuss Primitive was developed as a PoW substitute. While Proof of Stake (PoS) chooses the miner at random based on the stake that it has, Proof of Work (PoW) requires miners to solve a difficult mathematical puzzle in order to attain distributed consensus. The ether is locked while a block is being added to the network. Additionally, the locked Ether is freed after the block is successfully added. Any unauthorized effort to add a block may result in a penalty that is taken out of the already-locked Ether. Additionally, there isn't any rivalry among miners in PoS. It is practically extremely improbable for a block maker to possess more than 51% of the network's total cryptocurrency stake in order to create a fake block. PoS primarily saves energy by avoiding resources that need a lot of computation.

- PoA (Proof of Activity): PoA [9] is a hybrid strategy that uses distributed consensus to guarantee the legitimacy of the transactions and the achievement of common consensus. PoA is in the middle between PoW and PoS. Blocks are generated by the PoW mining method; however, in order to integrate the block into the network, a PoS strategy is employed, in which validators place bets in order to be chosen to mine the block. The advantages of PoW and PoS are indirectly transferred to PoA. An additional line of protection is added to the security of proof-of-work (PoA) by requiring the attacker to possess both (i) sufficient minted coins in terms of stake and (ii) strong computing mining power.
- PoL (Proof-of-Location): According to [4] PoL uses location verification, and the locations are block-coded. For location-dependent functions, PoL is helpful. The Proof of Location services rely on verified geospatial data, as well as open-source maps.
- PoI (Proof-of-Importance): According to [11], The Proof of Importance (PoI), developed by the NEM (New Economic Movement) group, determines which nodes are allowed to add block to the chain through a procedure known as mining. A node's likelihood of getting chosen to add a block will increase with its reputational importance score. For the node to be eligible for harvesting, it must own a minimum of 10,000 vested XEM.
- PoeT (Proof-of-Elapsed-Time): [9] introduced by Intel in 2016 and centered around a straightforward lottery system whose main goal is to provide each participating node in the network with an equal opportunity to add a block. Every node experience a random duration of sleep mode. The first node to wake up adds the block and becomes personal with the other nodes in the network [6].

11.3.3 Research challenges

We go over the main research issues and avenues that we think should be investigated in this part. [4] outlined the following seven technological obstacles and restrictions that will prevent Blockchain technology from being fully adopted in the future:

1. Throughput: This technology's throughput is indicated by the quantity of new transactions added within a predetermined time frame. Including the Bitcoin network increases the throughput to a maximum of 7 tps (transactions per second). In addition, blockchain technology must enhance its throughput capacity when compared to existing networks for transaction processing, such as Twitter and VISA, which have 5000 and 2000 transaction processing points, respectively.
2. Latency: The main focus of modern Blockchain technology is creating secure, unchangeable blocks. The majority of the time is devoted

to validation and verification to prevent fraudulent transactions and double spending. Due to security considerations, the construction of blocks and transaction confirmations take a long time. Thus, latency is a big problem for blockchain right now.

3. Size and bandwidth: The quantity of blocks generated determines a blockchain size. A block in Bitcoin is one megabyte in size and is generated every 10 minutes. As such, the maximum transactions that can be added to the block is limited.
4. Security: It is currently possible for blockchain technology to be attacked via a 51% attack. Often, the majority of the network can be fully controlled by a single entity. This is a challenge and possible security concern.
5. Wasted resources: Miners must perform a significant amount of compute-intensive mining labor in the permissionless environment. A lot of the mining work fails because of the consensus mechanism and time limits. As a result, mining resources and time are lost. To have more effective blockchain mining, this resource waste problem must be fixed.
6. Usability: User-friendly APIs are essential for blockchain applications. It has already been discovered that the Bitcoin API is challenging to utilize when creating new services. To increase Blockchain's appeal to developers, a more developer-friendly API must be created.

11.4 APPLICATION AREAS

In the quickly changing digital landscape, blockchain technology has revolutionized information exchange, transaction processing, and trust definition. Blockchain platforms record transactions and track assets, acting as a digital database or ledger system that promotes accountability and transparency within a business network (Figure 11.5).

A. Registration of Land: The process via which a buyer who want to purchase or is interested in purchasing land or a property uses is known as land registration. This system consists of a number of divisions, individuals, and intermediary third-party organizations. Land registration has grown difficult these days, therefore overcoming all of these difficulties Blockchain technology can be integrated with land registration. By utilizing blockchain technology, all current obstacles will be resolved. In land registration, for example, fraud situations can be overcome since, as this process goes digital, it will be more difficult to provide fraudulent documentation. Involving the buyer, seller, and land inspector is the main goal of blockchain. The user who want to purchase or sell real estate must register on the blockchain network. After creating a profile and providing government-approved

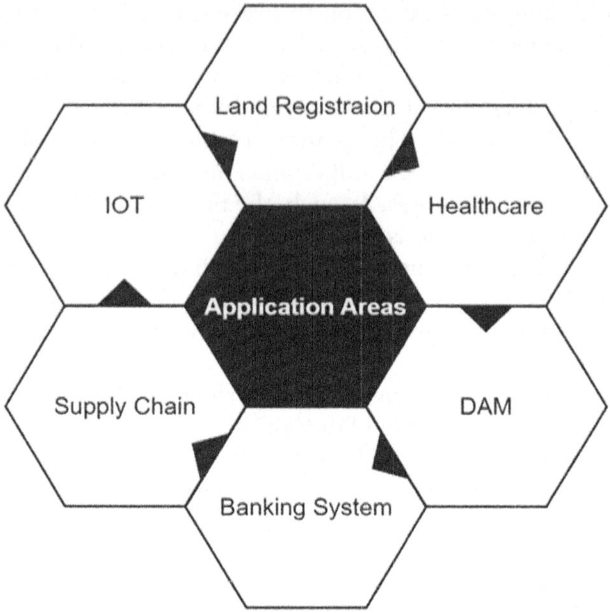

Figure 11.5 Application areas.

identification documents, the user's identity is hashed and uploaded to the blockchain [4].

B. Healthcare: Medical Services The healthcare sector is expanding daily due to new machines and technology that make it easier to treat terminal diseases. However, some nations fall short in this process, and even those that have complete medical facilities are lagging in some areas that they are unable to investigate, such as hospitals' inability to maintain records of the immunizations administered to patients and the patients' immunization due dates. In this location, there are also examples of fraud where people present fictitious vaccination certificates. However, tying the blockchain concept to the healthcare sector solves the problem the sector is facing. Operational value could be unlocked through a blockchain-powered health information exchange.
C. Banking: Financial System Technology and banking are closely related fields, and banking advances have evolved. when the timing is right. Everything has evolved, from the barter system to digital banking and digital signatures in place of wax seals. Blockchain technology is one of the newest and most disruptive innovations in the digital financial sector. Blockchain is a distributed ledger that records all of your transactions and makes them visible to everyone connected to the network. It stores transactions and allows anybody to view them. Blockchain technology can be used for trade finance, KYC, and digital banking. It gives

us a system that is adequately safe because all transactions are kept in a database with a hash value that is difficult to decrypt. As a result, polls claim that 51% of attacks target blockchain technology; nonetheless, when compared to other blockchains, this network is secure.

D. Supply Chain: Chain of Supply To put it more simply, supply chain management is the act of organizing and carrying out the movement of products, materials, and capital. The raw material is sent to the manufacturer, and the supply chain moves forward to provide the finished product to the customer. Although supply chain management can result in high sales and revenue, it takes a lot of work to keep up strong supply chain management. We can use blockchain technology to build a supply chain and address problems that the supply chain sector is facing. With a centralized database, it will be less prone to corruption and be able to quickly track records and products. Numerous elements, including fraud cases, inventory control systems, reducing courier expenses.
E. Digital Assets Management: Many issues with the traditional corporate digital asset management system, including copyright disputes, privacy disclosure, data loss, malicious manipulation, single point of failure, etc., can be resolved by an enterprise's blockchain-based digital asset management system. Blockchain technology, a distributed ledger, can lower the risk associated with centralized management by using techniques like multi-point maintenance, consensus mechanisms, encryption algorithms, and other technical measures to assure the veracity and trustworthiness of data on the chain.

11.5 USE CASE

11.5.1 Project overview

We are going to present a use case scenario of a healthcare system. In healthcare system several entities like doctor, labs, pharma, insurance company and patients are involved as shown in Figure 11.6. A patient goes through several stages in his/her treatment journey. Firstly, patient visits nearby hospital

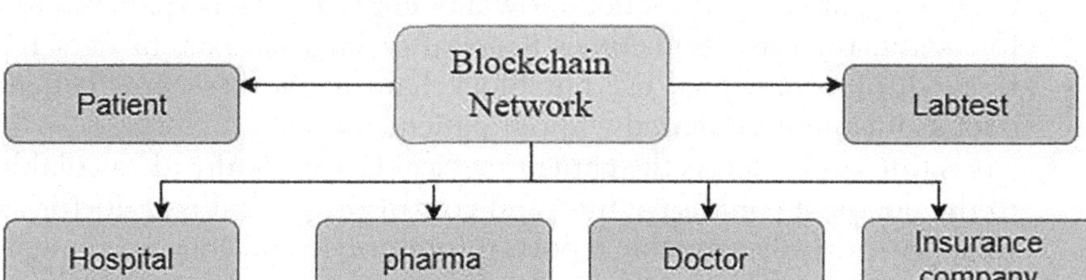

Figure 11.6 Entities involved in healthcare system.

and get himself registered and seeks for the doctor's appointment. In doctors' appointment doctor ask for several tests. then patient goes through these tests and shows these reports to the doctor. Doctor starts treatment based on reports and prescribe medicines to the patient. Patient purchase this medicine form pharma and follow the prescribed medicines for few days as recommended by the doctor.

In case if patient is recovered from the disease, he/she will stop medicine. In future if patient is facing any other medical issue or same issue he/she visits the doctor again. If all the entities are connected to the blockchain network. This time scenario would be bit different. medical history of the patient is available with doctor and all the medicines patient consumed in the past is recorded there. results of lab tests are also there on the network. Now doctor can start the treatment based on his/her medical history.

11.5.2 Advantages of blockchain in healthcare system

- Medical history of patient: In a Blockchain network patient treatment journey is always available to the doctor whenever and wherever a patient seeks doctor appointment. With the help of this historical data doctor can understand patient condition better and could start treatment immediately.
- Reports availability to doctor: patients previous lab report are available to the doctor and if a patient goes through some more tests again lab reports will be updated. It will help the doctor to understand patient history and sometime older reports can help doctor to prescribe the patient on that basis.
- Help insurance company for claim settlement: If a patient has medical insurance of a specific company and claim for medical bills. It will help the insurance company to track all the bills of medicines, lab test, doctors fee etc. this data is already available to the company and they do not need to verify as it cannot be changed by any party.
- Restrict data forging: Once data is updated on the blockchain network it cannot be altered. It will help every entity connected to healthcare system for trust building.
- Protect patient data: Blockchain network ensures security of data related to patient. It does not allow any entity to misuse patient data for commercial use. Sometimes hospital or labs sell patient data for research or commercial use but Blockchain network secures patient data as it cannot be shared without patient consent.
- Saves time and cost: As the patient medical history is already available to the doctor it could save time and cost of the patient and doctor as well. With already available reports or prescribed and consumed medicines doctor can treat patient immediately.

11.6 CONCLUSION

Blockchain technology has been studied in many fields since it was first introduced as a cryptocurrency, including supply chain management, finance, electronic voting, insurance, asset tracking, healthcare, the Internet of Things, and smart contracts. Through this thorough investigation, we have looked into a number of aspects of blockchain technology. However, because this technology is still in its early stages of development, many issues need to be extensively investigated before it can be widely utilized. Our analysis led us to identify and discuss some areas for further research in the different application sectors. The research community is anticipated to benefit from this study in understanding a number of areas, including future prospects for Blockchain technology and research problems [5].

REFERENCES

[1] J. Michael, A. Cohn, J. B.-T. Journal, and undefined 2018. Blockchain technology. *steptoe.com*, 2018, Accessed: Apr. 01, 2024. [Online]. Available: https://www.steptoe.com/a/web/171967/LITFebMar18-Feature-Blockchain.pdf

[2] M. V. the A. workshop on blockchain, undefined cryptocurrencies, and undefined 2017. Rethinking permissioned blockchains. *dl.acm.org M Vukolić Proceedings ACM Work. blockchain, cryptocurrencies Contract. 2017•dl.acm.org*, pp. 3–7, Apr. 2017, doi: 10.1145/3055518.3055526.

[3] S. Nakamoto 2008. Bitcoin: A peer-to-peer electronic cash system. Accessed: Apr. 01, 2024. [Online]. https://assets.pubpub.org/d8wct41f/31611263538139.pdf

[4] B. Shrimali and H. B. Patel 2021. Blockchain state-of-the-art: architecture, use cases, consensus, challenges and opportunities. *J. King Saud Univ. - Comput. Inf. Sci.* doi: 10.1016/j.jksuci.2021.08.005.

[5] S. Nakamoto 2008. Bitcoin: A peer-to-peer electronic cash system. Accessed: Apr. 04, 2024. [Online]. Available: https://assets.pubpub.org/d8wct41f/31611263538139.pdf

[6] H. Watanabe, S. Fujimura, A. Nakadaira, Y. Miyazaki, A. Akutsu, & J. J. Kishigami Feb. 2016. Blockchain contract: A complete consensus using blockchain. *2015 IEEE 4th Glob. Conf. Consum. Electron. GCCE 2015*, pp. 577–578. doi: 10.1109/GCCE.2015.7398721.

[7] Rana, S. K., & Rana, S. K. 2021. Intelligent amalgamation of blockchain technology with industry 4.0 to improve security. In *Internet of Things* (pp. 165–175). CRC Press.

[8] Z. Zheng, S. Xie, H. Dai, X. Chen, and H. Wang Sep. 2017. An overview of blockchain technology: Architecture, consensus, and future trends. *Proceedings - 2017 IEEE 6th International Congress on Big Data, BigData Congress 2017*, Institute of Electrical and Electronics Engineers Inc., pp. 557–564. doi: 10.1109/BigDataCongress.2017.85.

[9] L. M. Bach, B. Mihaljevic, & M. Zagar Jun. 2018. Comparative analysis of blockchain consensus algorithms *2018 41st Int. Conv. Inf. Commun. Technol. Electron. Microelectron. MIPRO 2018 - Proc.*, pp. 1545–1550. doi: 10.23919/MIPRO.2018.8400278.

[10] S. K. Rana, S. K. Rana, A. K. Rana, & S. M. Islam Nov. 2022. A blockchain supported model for secure exchange of land ownership: An innovative approach. *2022 International Conference on Computing, Communication, and Intelligent Systems (ICCCIS)*, pp. 484–489. IEEE.

[11] N. Bozic, G. Pujolle, & S. Secci Mar. 2017. A tutorial on blockchain and applications to secure network control-planes. *2016 3rd Smart Cloud Networks Syst. SCNS 2016*. doi: 10.1109/SCNS.2016.7870552.

[12] M. Gomez et al. 2019. Is blockchain the next step in the evolution chain of intermediaries? *Pap. Gomez, P Bustamante, MBH Weiss, I Murtazashvili, MJ Madison, W Law, T MylovanovTPRC47 47th Res. Conf. Commun. Inf. and ..., papers.ssrn.com*, Accessed: Apr. 04, 2024.

[13] S. K. Rana, A. K. Rana, & S. Dhawan 2022.A vital fusion of Internet of medical things and blockchain to transform data privacy and security. In *Convergence of Deep Learning and Artificial Intelligence in Internet of Things* (pp. 293–308). CRC Press.

[14] M. Castro, & B. Liskov, et al. 1999. Practical byzantine fault tolerance. *OSDI*, 99, 173–186. Accessed: Apr. 05, 2024.

[15] J. H. 2019. A lightweight BFT consensus protocol for blockchains. *arxiv.orgJ HackfeldarXiv Prepr. arXiv1903.11434, 2019.arxiv.org*, 2021, Accessed: Apr. 11, 2024.

[16] F. Armknecht, G. O. Karame, A. Mandal, F. Youssef, & E. Zenner 2015. Ripple: Overview and outlook. *Lect. Notes Comput. Sci. (including Subser. Lect. Notes Artif. Intell. Lect. Notes Bioinformatics)*, 9229, pp. 163–180. doi: 10.1007/978-3-319-22846-4_10.

Chapter 12

Advanced video steganography

Integrating edge detection with identical match approaches for optimal data hiding

Ramandeep Kaur, Inderpal Singh and Arun Rana

12.1 INTRODUCTION

12.1.1 Video steganography

Video steganography involves embedding secret messages within the frames of a video file, effectively masking confidential information. A video file, composed of a sequence of frames that combine audio and image files, can accommodate a substantial amount of hidden data. The creation of a video media file typically requires 16 fps, or 25 fps for high-definition (HD) videos. The primary objective of video steganography is to conceal data in such a way that it remains undetectable by the human visual system (HVS), without degrading the video's quality [1-4]. Various formats, such as AVI, FLV, MP4, MOV, and H.264, are utilized in the steganography process. The fundamental equation for video steganography is expressed as follows:

$$\text{Cover\_Video} + \text{Secret\_Msg} + \text{Secret\_Key} = \text{Stego\_Video} \quad (12.1)$$

12.1.2 Basic process of video steganography

Video steganography operates through two primary phases: the embedding and the extraction phase. During the embedding phase, the confidential message is hidden within the video file using specific algorithms, typically conducted on the sender's side. Conversely, the extraction phase, which is the reverse process of the embedding process, involves retrieving the hidden message from the video file at the receiver's end.

12.1.2.1 Elements of video steganography

The video steganography process comprises five critical components:

1. *Cover-Video (C)*: The original video that serves as the carrier for secret information.
2. *Secret Message (M)*: The actual information embedded within the cover video, which could be text, image, audio, or video.

DOI: 10.1201/9781003590439-12

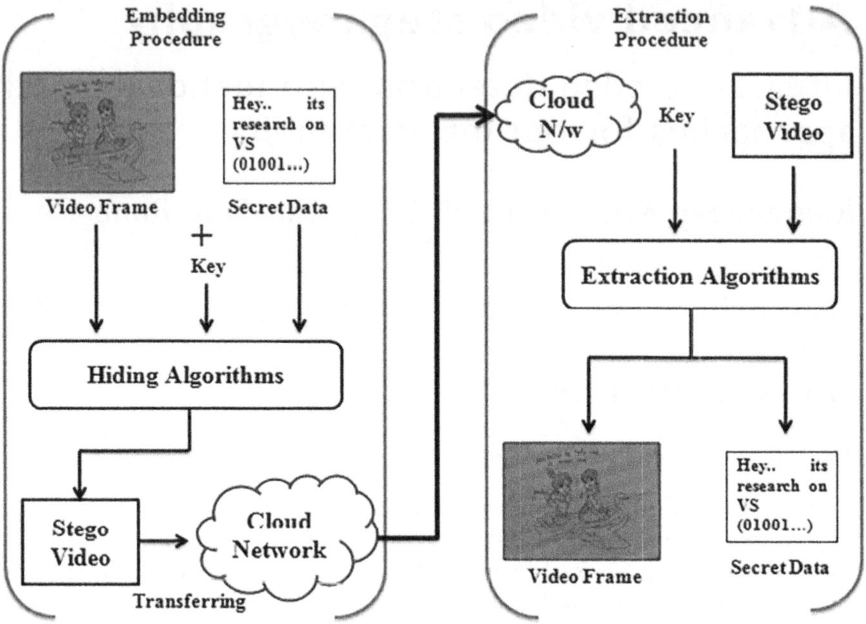

Figure 12.1 Process of video steganography.

3. ***Stego-Video (S)***: The resultant video file after the secret message has been embedded, now containing the hidden message.
4. ***Stego-Key (K)***: A key utilized for embedding or extracting secret messages from the cover-video and stego-video respectively (Figure 12.1).

12.1.3 Video steganalysis

Steganalysis is the technique used to analyze the presence of hidden messages within media files, such as images and videos. It is crucial for cybersecurity, aiding in the prevention of cybercrimes and tracking cybercriminals, including hackers and data thieves. Steganalysis can determine whether incoming media is genuine or fraudulent and identify malicious users by analyzing various parameters such as contrast, dimensions, and histogram analysis of the secret message before and after the hiding process. Key steganalysis techniques include:

a. **Histogram Analysis**: This technique identifies significant changes in color frequency values by comparing the cover file to the stego file, focusing on attributes such as brightness, sharpness, and RGB color contrast.
b. **Authentication Check**: This involves identifying a stego video using labeling information, such as password verification or watermark analysis. Unauthorized users cannot extract the hidden message.

c. **Attacks:** Various attacks, including visual, statistical, and structural, are employed to breach the security of stego media files by comparing the original and modified versions.

12.2 LITERATURE REVIEW

12.2.1 Hemant Gupta and Setu Chaturvedi (2014)

In their paper titled "Video Steganography Through LSB Based Hybrid Approach" published in the International Journal of Computer Science and Network Security (Vol. 14, March 2014) [5], the authors propose a hybrid video steganography technique combining 1 LSB, 2 LSB, and 3 LSB substitution methods with AES encryption. Their findings indicate that as the number of LSB bits increases, the PSNR decreases. Thus, the 3 LSB substitution technique offers enhanced security for hiding secret bits compared to 1 LSB and 2 LSB methods. Additionally, this approach yields a high correlation factor (r) between the cover and stego frames, where r = 1 indicates good correlation, r = 0 indicates no correlation, and r = –1 signifies anti-correlation, meaning one frame is the negative of the other.

12.2.2 Arijit Basu, Gaurav Kumar, and Soumyajit Sarkar (2014)

Their study, "A Video Steganography Approach using Random Least Significant Bit Algorithm," published in the International Journal of Science and Research (Vol. 3, Issue 6, June 2014) [6], introduces a novel steganography method utilizing random LSB bits to embed secret message bits within video files. This approach achieves high PSNR, low Mean Squared Error (MSE), and superior quality stego videos. The results demonstrate that the Random LSB (RLSB) technique provides a higher PSNR compared to traditional LSB methods.

12.2.3 Krishna Nand Chaturvedi and Amit Doeger (2014)

In their paper, "A Novel Approach for Data Hiding using LSB on Edges of a Gray Scale Cover Images," featured in the International Journal of Computer Applications (Vol. 86, Issue 7, January 2014) [7], the authors propose an innovative data hiding methodology using LSB substitution on edge pixels of grayscale images rather than on smooth areas. This edge-based steganography embeds secret bits within the edge regions of grayscale images using the Canny edge detector and 2 LSB techniques. The study concludes that the Enhanced LSB (ELSB) method is more effective than traditional LSB substitution, achieving high PSNR values.

12.2.4 Soumyajit Sarkar and Arijit Basu (2014)

Their research, titled "Comparison of Various Edge Detection Techniques for Maximum Data Hiding Using the LSB Algorithm," was published in the International Journal of Computer Science and Information Technologies (Vol. 5, Issue 3, 2014) [8]. The study compares different edge detection techniques to optimize data hiding using the LSB algorithm. It evaluates edge detectors such as Sobel, Prewitt, Roberts, Canny, and Laplacian based on PSNR and MSE quality metrics. The experimental results indicate that the Canny edge detector is the most effective operator for detecting true edge pixels in images, offering high capacity rates and low MSE values, despite having lower PSNR values compared to other detectors.

These studies collectively enhance the understanding of video steganography and contribute to the development of more secure and effective data hiding techniques.

12.3 PROPOSED METHODOLOGY

This research consists of five main phases, outlined in the flowchart in Figure 12.2.

i. **Pre-processing**: Initial loading and frame calculation of a video clip (.avi format).
ii. **Frame Fragmentation and Random Frame Selection**: Conversion of video to frames and selection of random frames using a secret key to enhance security.
iii. **Edge Detection**: Detection of edges in selected frames using the Canny edge detector, which is optimal for identifying true edge pixels and reducing noise.
iv. **RSA Encryption**: Conversion of the confidential text message into an unreadable format using the RSA algorithm to provide additional security.
v. **Embedding Process**: Embedding of secret data into edge and non-edge pixels using four-LSB and 7 Pair Identical Match techniques to maximize data capacity and video quality.
vi. **Extraction Process**: Retrieval of hidden data from the stego video at the receiver's end using the reverse algorithm of the embedding process.

12.3.1 Phase I: pre-processing

The pre processing involves initial phase of thesis work in which a video clip of format (.avi) is loaded and total number of frames are calculated. In pre-processing two steps are used which are explained as follow:-

a. ***Frame Fragmentation***: The loaded video clip is converted into frames to extract total number of frames in a video file using video to frame conversion functions. These frames are used for hiding the high amount of data without degrading the quality of each frame.

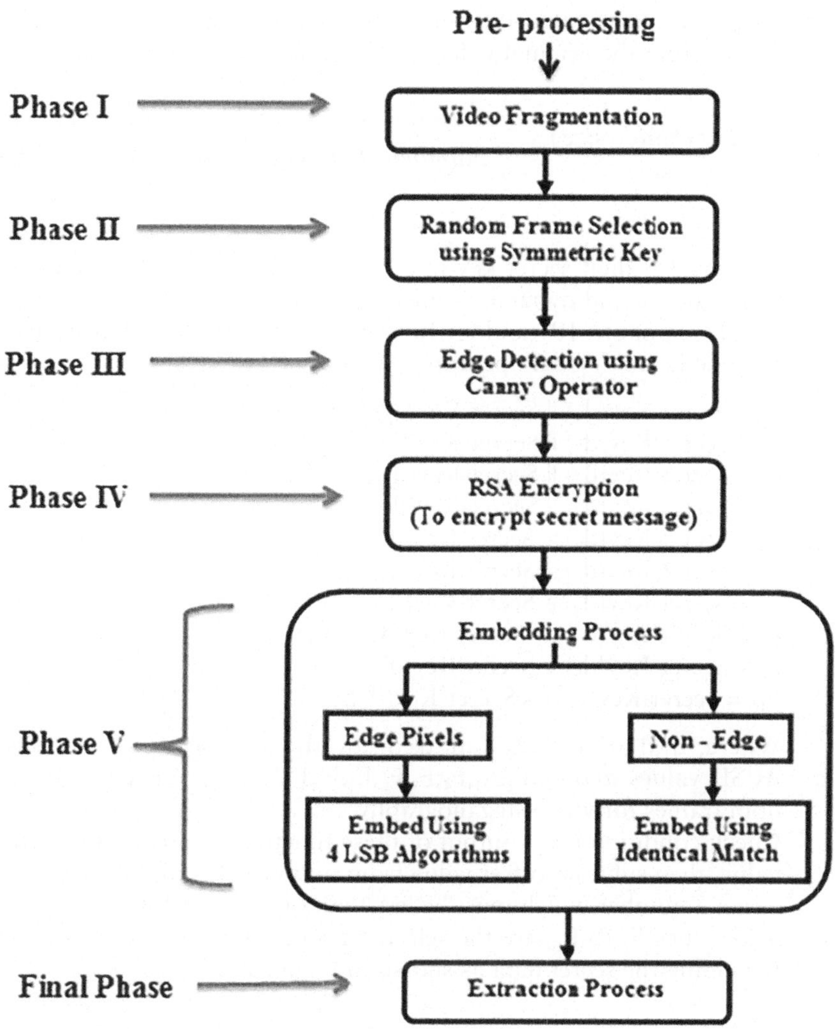

Figure 12.2 Flow chart of proposed methodology.

b. *Random Frame Selection*: Instead of using sequential frames, we used some random frames for hiding the secret message to improve the security of hidden data using secret key and by creating some random functions for selecting the frames from all video frames and avoid problem like frame dropping attack. The algorithm for edge detection is explained as follows.

1. Select video clip (.avi) and calculate total number of frames (nof).
2. Covert video into frames using video2frame function.
3. Enter 10 digit secret key for providing authentication to video clip.
4. Select 10 random frames from all frames of video by using secret key.

5. Calculate offset value (Slab-Size) for selecting the any ten random frames from the original video file by using following formula:

$$Offset\left(Slab-size\right)=\frac{Total\ number\ of\ frames\ in\ video\left(nof\right)}{Number\ of\ frames\ want\ to\ select}=\frac{nof}{10} \quad (12.2)$$

6. Convert 12- digit secret key into ASCII code values for selecting the then random and dynamic frames from video for steganography process & compute 10 random functions for frame selection by using arithmetic and logical operations such as [4]:

   F_1 = Secret-Key(d_1) || Secret-Key(d_2);
   F_2 = Secret-Key(d_3) || Secret-Key(d_4);
   F_3 = Secret-Key(d_4) || Secret-Key(d_5);
   F_4 = Secret-Key(d_6) || Secret-Key(d_7);
   F_5 = Secret-Key(d_8) + Secret-Key(d_5);
   F_6 = Secret-Key(d_9) + Secret-Key(d_6);
   F_7 = Secret-Key(d_8) + Secret-Key(d_9);
   F_8 = Secret-Key(d_{12}) + Secret-Key(d_2);
   F_9 = Secret-Key(d_9) + Secret-Key(d_{10});
   F_{10} = Secret-Key(d_{11}) + Secret-Key(d_{12});

   Where, $d_1,d_2,d_3,d_4,d_5,d_6,d_7,d_8,d_9,d_{10},d_{11},d_{12}$ are the secret key digit's ACSII values and $F_1,F_2,F_3,F_4,F_5,F_6,F_7,F_8,F_9,F_{10}$ represent the 10 random frames for video steganography.
7. Check condition i.e. if output of these functions is greater than offset value, then subtract offset value from output value of these functions to get a number which must be less than offset or slab-size.
8. The values F_1 to F_{10} are the selected frame addresses which are used for hiding the secret data as shown in Figure 12.3.

Figure 12.3 Showing selected 10 random frames for AVI Video ("Rhinos.avi").

12.3.2 Phase 2: edge detection

The selected random frames are used for the next phase of the thesis. In this phase, the edges of each random frame are detected using the Canny edge detector. The Canny edge detector is an optimal operator that accurately detects a greater number of true edge pixels by employing high and low threshold values. Additionally, it provides noise-free edges due to the Gaussian filter and ensures a single response to each edge. Edge pixels are inherently sharp, with intensity values that may vary in brightness. Consequently, when data is hidden within edge boundaries, the human eye cannot easily detect minor variations in color due to their sensitivity [9-14]. To improve the quality we are detecting edge frames to hide data. The algorithm for edge detection is explained as follows.

1. Select random frames and decompose them into R, G, and B layers to detect edges of each channel for all frames.
2. Apply canny edge detector on each channel for all selected frames and detect edge pixels of each layer such as edge-R, edge-G, & edge-B as shown in Figure 12.4.
3. Recombine all layers using OR operator (|) and generate edged image of all selected frames.

12.3.3 Phase 3: RSA encryption

In this phase, user enters secret text message of any length and then this text message is converted into unreadable format using RSA algorithm to make it more secure from intruders. RSA is asymmetric key based cryptographic algorithm which provides more security to secret data due to factorization of large prime numbers and creates difficulty for attackers to manipulate or decrypt the data. It avoids social media problems such as data manipulation [15-21].

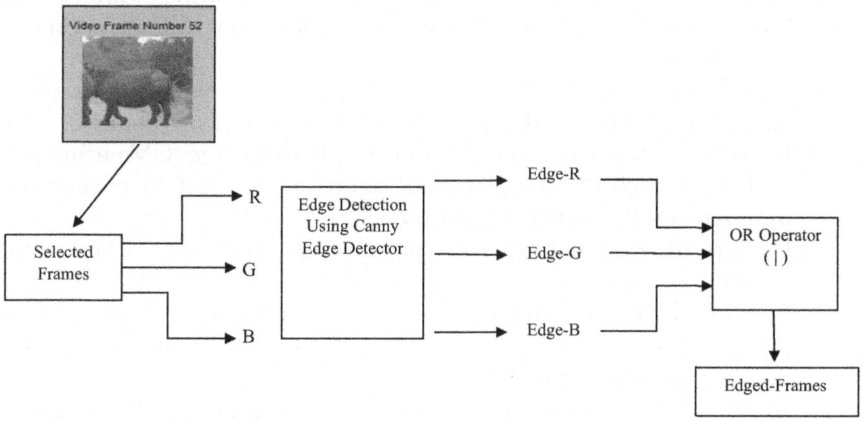

Figure 12.4 Showing basic process of edge detection algorithm.

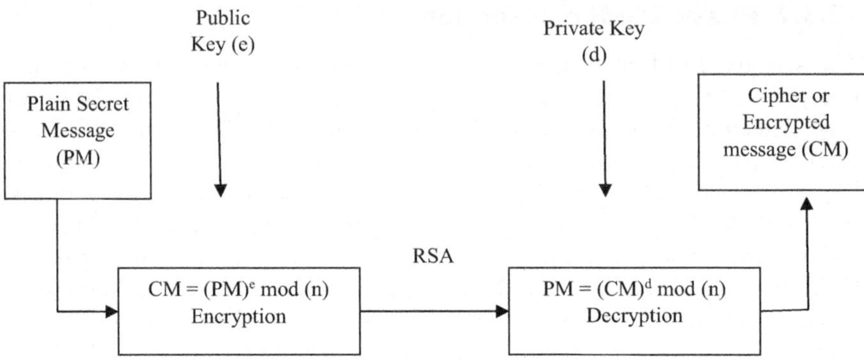

Figure 12.5 Showing implementation of RSA encryption.

The RSA encryption implementation is described as follows and shown in Figure 12.5 [22].

1. Input a text secret message (.txt) and encrypt it using the RSA encryption algorithm.
2. Select two large prime numbers, ppp and qqq, each divisible only by 1 and itself.)
3. Multiplies p & q to find modulus of encryption and decryption (n) i.e.

$$\text{Mod-n}(n) = p^{*}q \tag{12.3}$$

4. Calculate phi (Φ), which is used for calculating public and private keys i.e.

$$\text{Phi-m}(\Phi) = (p\text{-}1) * (q\text{-}1) \tag{12.4}$$

5. Choose (e) a small number which will be the co-prime of phi (Φ) i.e. Pub-e (e) = GCD (e, Φ). Where, e is a public key for encrypting the secret message.
6. Find d, such that (d*e)%Φ = 1 i.e. d*e = (1+nm)/e, where n any integer value that satisfy the above condition. Here, d is a private key.
7. Encrypt the plain message (PM) into cipher message (CM) using public key (e) such that, CM = $(PM)^e$ mod (n), where CM is encrypted message and PM is Plain text message.
8. Decrypt the encrypted message using private key (d) i.e. PM = $(CM)^d$ mod(n).
9. In RSA Algorithm, sender announces the public key (e), prime numbers (p, q) and n but keep private key (d) and phi (Φ) as a secret. Only receiver knows the private key.
10. After encrypting message, we get an encrypted message (CM) and we hide it behind video frames.

12.3.4 Phase 4: embedding process

In the embedding process, we conceal secret data within video frames. The primary goal of this phase is to enhance both data capacity and video quality. In our approach, we embed secret data bits in both edge and non-edge pixels [23-29]. Two fundamental substitution techniques are utilized for embedding the secret data: the 4 LSB substitution and the 7-pair Identical Match approach, as illustrated in Figure 12.6 and explained below:

a. ***4 LSB Substitution Approach***: For edge pixels, 4LSB technique is used for hiding 4 bits of secret message. LSB is a least significant bits based basic substitution method for steganography process but to increase the capacity of data we are hiding 4 bits of message in a particular edge frame pixels which is known as 4LSB technique. In 4LSB, the 4 least significant bits of a pixel in edge frame are directly replaced with 4 secret bits of message [15].
b. ***7 Pair Identical Match Approach***: For non- edge pixels, a novel embedding technique is used for hiding 4 secret bits. In 7 pair identical match (ID), instead of directly replacing the secret bits with 4 LSB bits of video frame we firstly makes 7 pairs of bits in a pixel of video frame and then compare these pairs with secret message bits [30]. If there is existence of identical pair then we jot down bit position of that pair in a table. It increases capacity, reduces resolution change in frames and also increases quality [31].

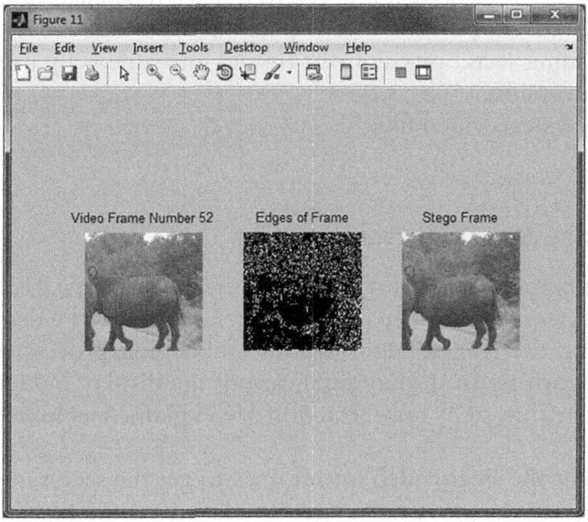

Figure 12.6 Showing original, edged & stego frames.

The algorithm for embedding process in edge pixels and non-edge pixels is as follows:-

1. Embed the secret message length, selected frame addresses, and secret key into the first frame of the video.
2. Encrypt and embed the message: Embed the encrypted message in edge pixels using the 4 LSB approach. And Embed the encrypted message in non-edge pixels using the 7-pair identical match approach.
3. The steps for embedding data on edge pixels are as below:
   - Check selected pixels of selected frames are edge pixels or not. If (Edge-Frame (i, j) = = 1), than edge exists. Else, Non- Edge pixels where, i and j are representing the row and column number of selected pixel of frame [15].
   - If condition is successful, than apply 4LSB substitution to hide 4 secret bits in blue layer of all selected frames.
   - Repeat step (b) for all pixels of all frames.
   - Repeat this algorithm for hiding all encrypted message bits.
4. The steps for embedding data on non-edge pixels are as follow:
   - Check selected pixel (Edge-Frame (i, j) = = 1). If no, then selected pixel of selected frame is a non- edge pixel where, I and j are representing the row and column number of selected pixel of frame.
   - Now, Apply 7 pair - Identical match technique to hide secret bits (2-bits in red layer and 2-bits in green layer of each frame) as shown in Figure 12.6.
   - Hide 2 bits of message in each place of pixel of selected frame by identifying the identical pairs of bits. I.e. If (Selected Pixel 2-bits = = 2- message bits), then no need to replace the bits as in LSB. Just note down the bit position of that pixel in binary table [31] (Figure 12.7).
5. After hiding secret bits, convert frames into video clip using frame2video conversion.
6. Generate stego video file.
7. Exit.

12.3.5 Phase 5: extraction process

The extraction operation is carried out on the receiver side. Using extraction process, we can extract hidden data from generated stego video clip [31-38]. By performing the reverse algorithm of embedding process, we create an extraction algorithm to display final output i.e. display hidden secret message. The algorithm steps for extraction are explained as follows.

1. Fragment the stego video into frames to get the secret message.
2. Decompose the red, green, and blue layers of the first fixed frame of the Stego video to get selected frame addresses, secret key, and message length.

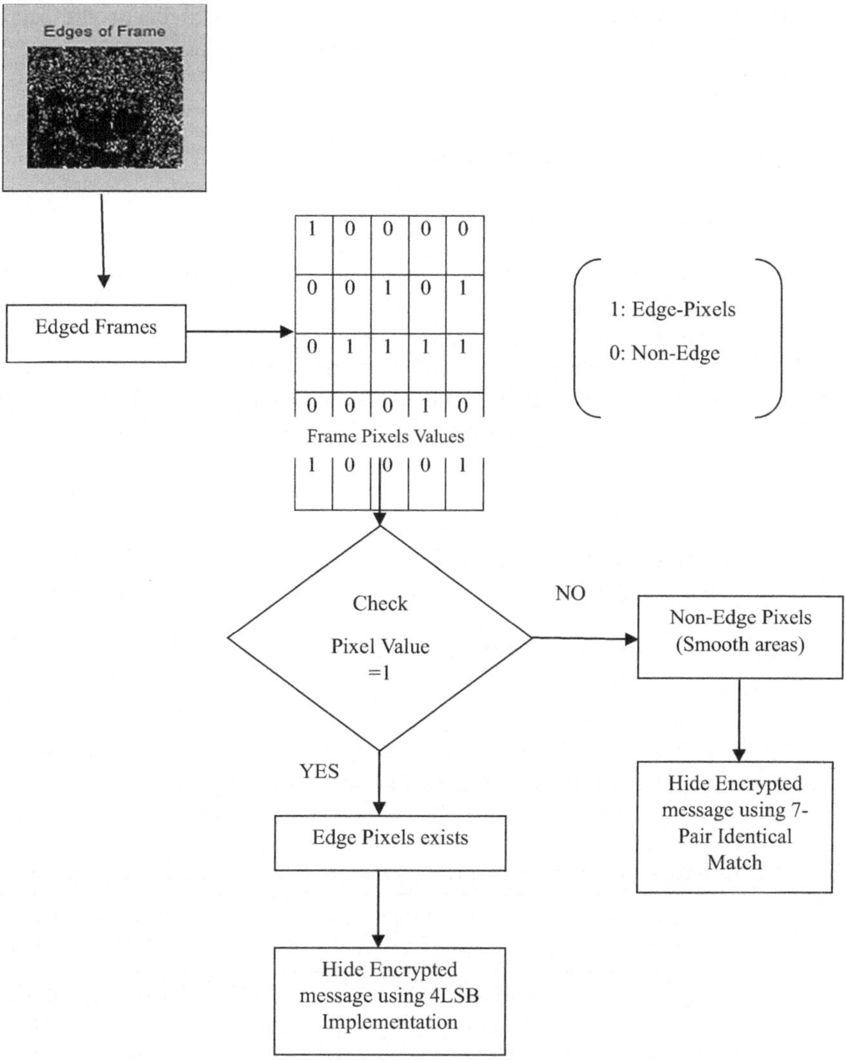

Figure 12.7 Flow chart for methodologies of data hiding.

3. Obtain the user's secret key and compare it to the original key entered or extracted during authentication. If successful, then continue with the extraction process. Otherwise, the video will be damaged due to unlawful access after four attempts.
4. Extract encrypted data from stego video frames using selected frame addresses and by performing reverse embedding algorithm.
5. Take private key (d) from user to decrypt the encrypted message and check it with originally computed value of private key.

6. If private key is correctly entered, then user will decrypt the secret message as final output of video steganography process.
7. Exit.

12.4 RESULTS AND DISCUSSION

12.4.1 Quality measurements

In research, four major performance measurements are used to determine final result values. PSNR, or peak signal to noise ratio, is used to assess quality. The errors are quantified using the mean error square (MSE) and bit error rate (BER). For a good steganography, PSNR value should be high and MSE & BER must be low. Other measurements are data hiding capacity and histogram analysis. All the measurements are explained as follows with their equation.

12.4.1.1 Payload capacity

The maximum data that can be hidden within the video frames, calculated using a specific equation and compared between previous and proposed methods.

$$\text{Capacity} = \left(\text{Total no. of rows m}^{*} \text{ cols n}^{*} \text{ total no. of bits hide in a pixel}\right) \tag{12.5}$$

Table 12.1 shows the data hiding capacity and test analysis for hiding different amount of data in same video. The table shows comparison between

Table 12.1 Comparison of previous & proposed work on basis of data hiding capacity

Name of video	*Hiding capacity in bytes*	*Previous PSNR (dB)*	*Proposed PSNR (dB)*	*Previous BER*	*Proposed BER*
Rhinos.avi	128 bytes	78.44	96.98	0.0127	0.0104
	256 bytes	76.31	90.61	0.0131	0.0118
	512 bytes	73.23	85.17	0.0137	0.0118
	1024 bytes (1KB)	70.26	80.22	0.0142	0.0126
	2048 bytes (2 KB)	67.19	75.11	0.0149	0.0134
	4096 bytes (4 KB)	64.17	71.65	0.0156	0.0141
	8192 bytes (8KB)	61.16	68.04	0.0164	0.0149
	16384 bytes (16KB)	58.15	63.45	0.0172	0.0159
	32768 bytes (32KB)	55.15	57.92	0.0181	0.0173
	65536 bytes (64KB)	ERROR: Payload Capacity Exceeded to its maximum limit i.e. 32 KB	53.55	-	0.018

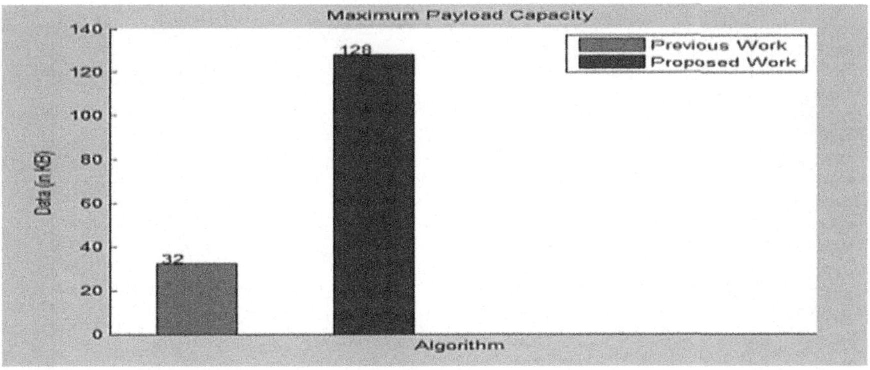

Graph 12.1 Comparison graph of data hiding capacity.

previous and proposed work on the basis of capacity i.e. from 128 bytes to 64 KB data.

Graph 12.1 compares the maximum data capacity of prior and proposed work, and our approach achieves a higher capacity (128 KB) than the previous one.

12.4.1.2 Peak Signal to Noise Ratio (PSNR)

Determines the quality of stego frames by comparing the peak signal to noise ratio between original and stego video frames. The higher the PSNR, the better the quality of the reconstructed stego frames, and the PSNR findings for the previous and suggested algorithms are computed using the formula shown below:

$$\text{PSNR} = 10^{*}\ \log_{10}\left(\frac{255^{*}255}{\text{MSE}}\right) \tag{12.6}$$

The following Graph 12.2 and 12.3, shows the comparison between previous and proposed work and proposed work is achieving high PSNR up to 76.57 dB than previous work i.e. 67.11 dB and generate high quality stego videos after hiding large amount data behind random video frames.

12.4.1.3 Bit Error Rate (BER)

Assesses the error between two image frames, with a lower BER indicating better steganography quality.

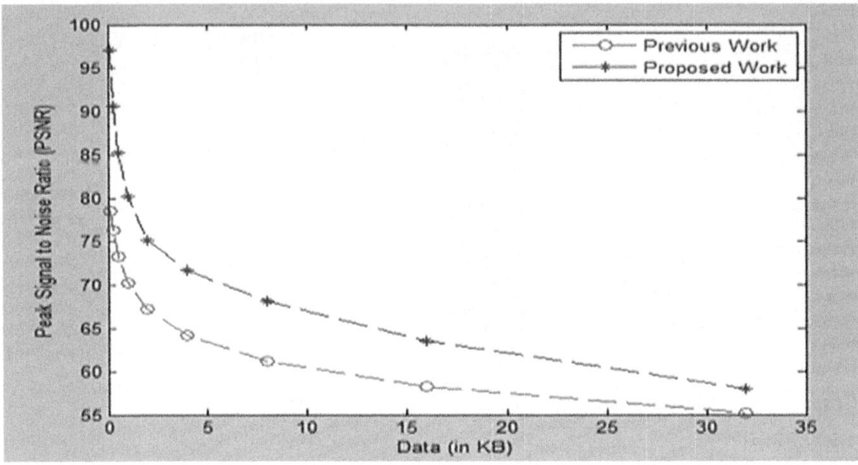

Graph 12.2 PSNR Comparison Graph for Different Amount of Data.

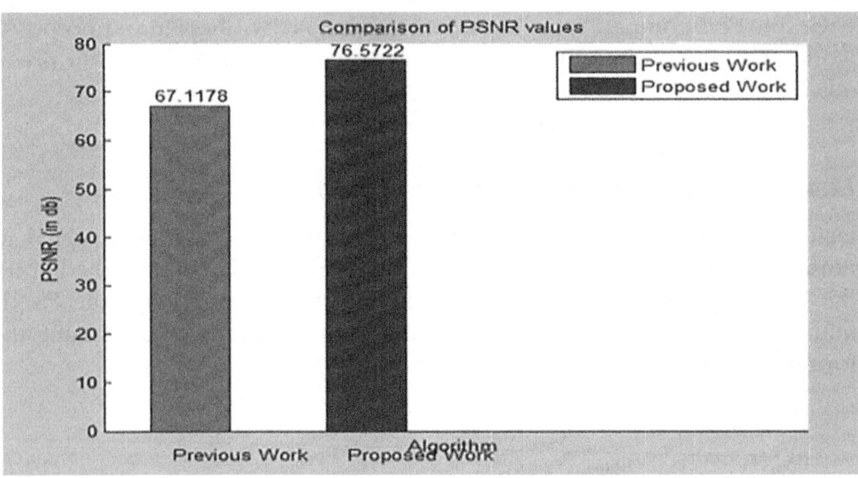

Graph 12.3 Average value of PSNR for Previous & Proposed Work.

It is reciprocal of PSNR. The basic equation is used for calculating bit error rate is showing as follows.

$$BER - \frac{\text{Number of Errors}}{\text{No. of Bits sent}} \tag{12.7}$$

The following Graph 12.4 and 12.5, shows the comparison between previous and proposed work on basis of bit error rate and proposed work is achieving low bit error rate of value 0.0130 than previous work i.e. 0.0147.

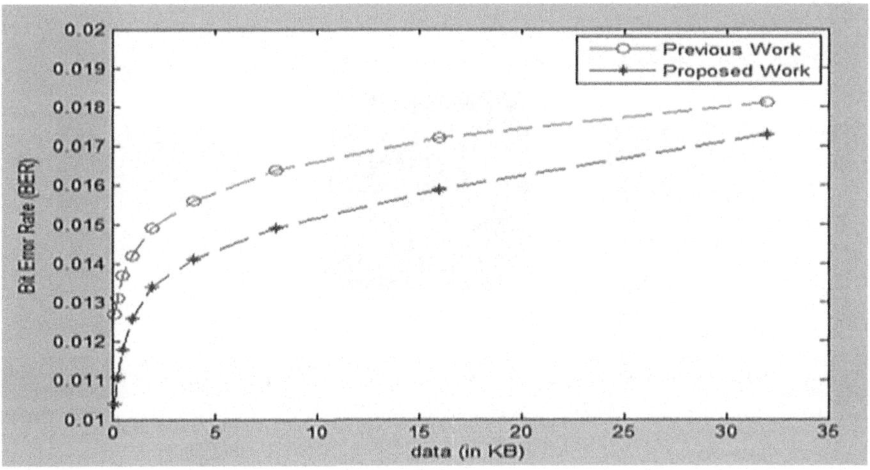

Graph 12.4 Comparison Graph for BER.

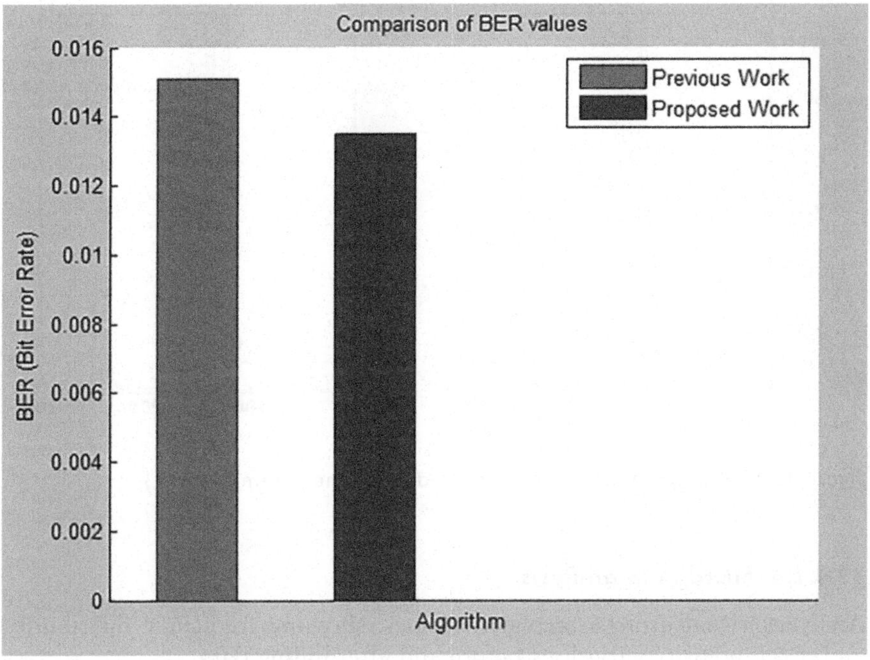

Graph 12.5 Average Value of BER for Previous & Proposed Work.

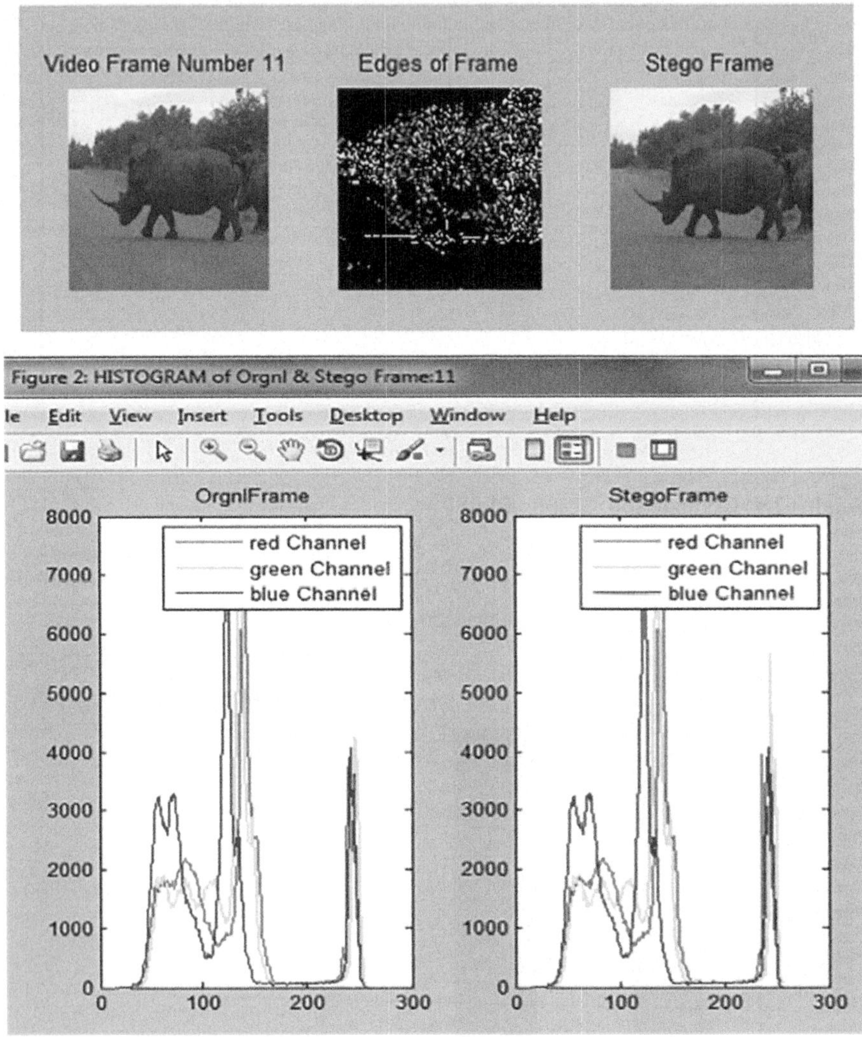

Figure 12.8 Histogram analysis for a random frame (Frame No.11).

12.4.1.4 Histogram analysis

Analyzes visual attacks on video frames, showing frequency distribution and pixel intensity variations before and after hiding data.

Figures 12.8 and 12.9 are showing the histogram analysis for some random video frames and analysis show that there is no difference in the frequency distribution corresponding to intensity pixel values in original and stego video frames.

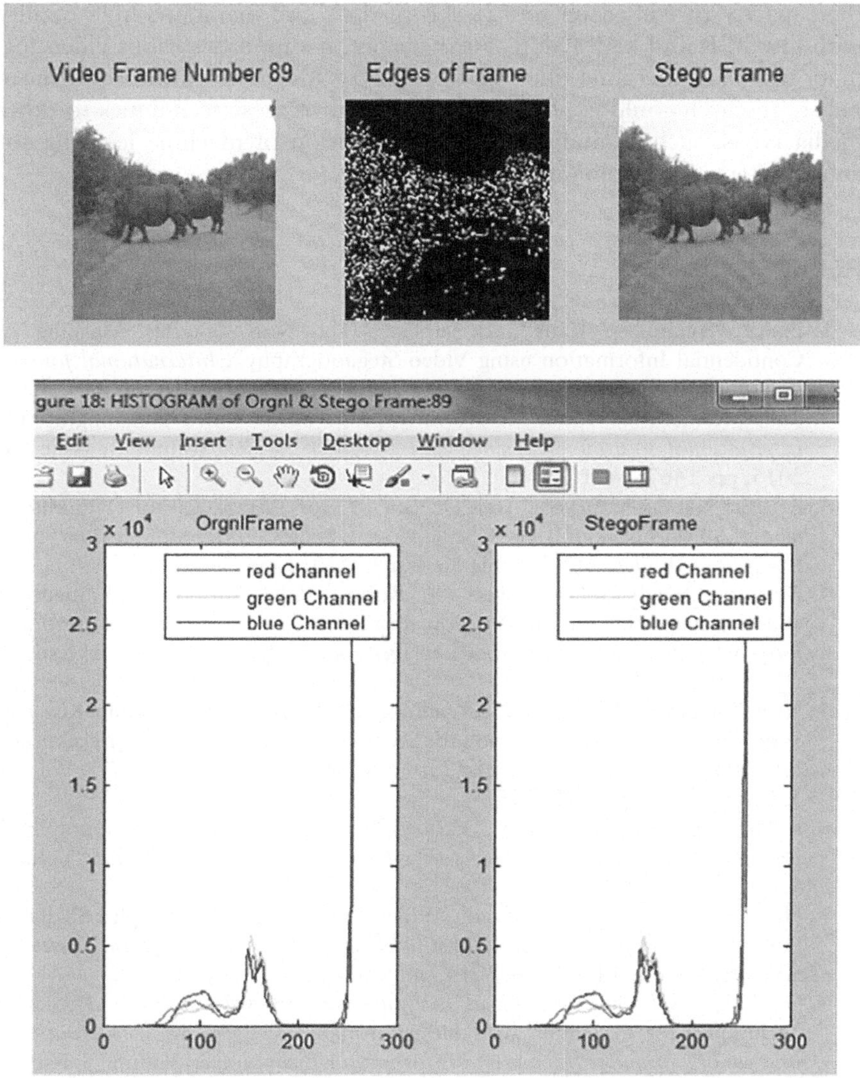

Figure 12.9 Histogram analysis for another random frame random frame (Frame No. 89).

12.5 CONCLUSION AND FUTURE SCOPE

12.5.1 Conclusion

This research presents a hybrid approach to video steganography, demonstrating significant improvements over existing techniques in terms of data capacity, security, and video quality. The proposed methodology enhances the capacity to embed up to 128 KB of data and maintains high quality with low BER and high PSNR. Furthermore, it supports various video formats and has undergone rigorous steganalysis testing to ensure robustness. Future research could explore the application of these techniques to other media types, such as audio, and the integration of machine learning for enhanced steganography processes.

REFERENCES

1. Pooja Ramandeep Kaur and Varsha, "The Non Tangible Masking of Confidential Information using Video Steganography", *International Journal of Computer Applications (IJCA)*, Volume 119, Issue 17, 2015, pp. 19–26.
2. Pooja Ramandeep Kaur, "XOR Encryption Based Video Steganography", *International Journal of Science and Research (IJSR)*, Volume 4, Issue 11, 2015, pp. 1467–1471.
3. Krishna Nand Chaturvedi, Amit Doeger, "A Novel Approach for Data Hiding using LSB on Edges of a Gray Scale Cover Images", *International Journal of Computer Applications*, Volume 86, Issue 7, 2014, pp. 36–40.
4. Amrinder Singh, Sukhjit Singh, "A Robust Video Watermark Embedding and Extraction Technique Based on Random Frame Selection" *International Journal of Research in Information Technology (IJRIT)*, Volume 2, Issue 2, 2014, pp. 28–37.
5. Nithiyanantham Chandran, S. Arulkumar, "Utilization of Random Key and Sobel Filter based Edge Detection for Secure Data Transmission", *International Journal of Computer Science and Information Technologies (IJCSIT)*, Volume 5, Issue 1, 2014, pp. 123–125.
6. K. Parvathi Divya, K. Mahesh, "Various Techniques in Video Steganography-A Review", *International Journal of Computer & Organization Trends*, Volume 5, 2014, pp. 8–10.
7. Hemant Gupta, Setu Chaturvedi, "Video Steganography Through LSB Based Hybrid Approach", *International Journal of Computer Science and Network Security*, Volume 14, Issue 3, 2014, pp. 99–106.
8. Soumyajit Sarkar, Arijit Basu, "Comparison of various Edge Detection Techniques for maximum data hiding using LSB Algorithm", *International Journal of Computer Science and Information Technologies*, Volume 5, Issue 3, 2014, pp. 4722–4727.
9. Shivani Khosla, Paramjeet Kaur, "Secure Data Hiding Technique Using Video Steganography and Watermarking", *International Journal of Computer Applications*, Volume 95, Issue 20, 2014, pp. 7–12.
10. Saurabh Singh, Ashutosh Datar, "Improved Hash Based Approach for Secure Color Image Steganography using Canny Edge Detection Method", *International Journal Computer Science and Network Security*, Volume 14, Issue 7, 2014, pp. 82–89.
11. Arijit Basu, Gaurav Kumar, Soumyajit Sarkar, "A Video Steganography Approach using Random Least Significant Bit Algorithm", *International Journal of Science and Research (IJSR)*, Volume 3, Issue 6, 2014.

12. P. Parveen, R. Arun, "Audio-video Crypto Steganography using LSB substitution and advanced chaotic algorithm", *International Journal of Engineering Inventions*, Volume 4, Issue 2, 2014, pp. 01–07.
13. Mamta Juneja and Parvinder Singh Sandhu, "Improved LSB based Steganography Techniques for Color Images in Spatial Domain", *International Journal of Network Security*, Volume 16, Issue 6, 2014, pp. 452–462.
14. Jaspreet Kaur, Gurbinder Singh Brar, Rahul Malhotra, "A steganography Technique on Edges", *International Journal of Engineering Research & Technology (IJERT)*, Volume 3, Issue 11, 2014, pp. 1408–1411.
15. Sunil K. Moon, Rajeshree D. Raut, (2013), "Analysis of Secured Video Steganography Using Computer Forensics Technique for Enhance Data Security", *IEEE Second International Conference on image information processing (ICIIP-2013)*, pp. 660–665.
16. Pooja Yadav, Nishchol Mishra, and Sanjeev Sharma, "A Secure Video Steganography with Encryption Based on LSB Technique", *IEEE International Conference on Computational Intelligence and Computing Research*, 2013.
17. Sarabjeet Kaur and Sonika Jindal, "Image Steganography using Hybrid Edge Detection and First Component Alteration Technique", *International Journal of Hybrid Information Technology*, Volume 6, Issue 5, 2013, pp. 59–66.
18. Sneha Arora, Sanyam Anand, "A Proposed Method for Image Steganography Using Edge Detection", *International Journal of Engineering Science*, Volume 8, Issue 2, 2013, pp. 22–25.
19. Mehdi Hussain and Mureed Hussain, "A Survey of Image Steganography Techniques", *International Journal of Advanced Science and Technology*, Volume 5, 2013, pp. 113–124.
20. Mamta Juneja, Parvinder Singh Sandhu, "A New Approach for Information Hiding in Color Images using Adaptive Steganography and Hybrid Feature Detection with Improved PSNR and Capacity", *International Journal of Engineering and Technology (IJET)*, Volume 5, Issue 2, 2013, pp. 1853–1862.
21. Vipul Sharma, Sunny Kumar, "A New Approach to Hide Text in Images Using Steganography", *International Journal of Advanced Research in Computer Science and Software Engineering*, Volume 3, Issue 4, 2013, pp. 701–708.
22. Anil Kumar, Rohini Sharma, "A Secure Image Steganography Based on RSA Algorithm and Hash-LSB Technique", *International Journal of Advanced Research in Computer Science and Software Engineering*, Volume 3, Issue 7, 2013, pp. 363–372.
23. K.V. Vinodkumar, V. Lokeswara Reddy, "A Novel Data Embedding Technique for Hiding Text in Video File using Steganography", *International Journal of Computer Applications*, Volume 77, Issue 17, 2013, pp. 13–18.
24. Er Prajna Vasudev, Er Kumar Saurabh, "Video Steganography Using 32 *32 Vector Quantization of DCT", *International Journal of Software & Hardware Research in Engineering*, Volume 11, Issue 3, 2013, pp. 40–43.
25. Poonam Dhankhar, Neha Sahu, "A Review and Research of Edge Detection Techniques for Image segmentation", *International Journal of Computer Science and Mobile Computing*, Volume 2, Issue 7, 2013, pp. 86–92.
26. S. Vijayarani, M. Vinupriya, "Performance Analysis of canny and Sobel Edge Detection Algorithm in Image Mining", *International Journal of Innovative Research in Computer and Communication Engineering*, Volume 1, Issue 8, 2013, pp. 1760–1767.

27. Babloo Saha and Shuchi Sharma, "Steganographic Techniques of Data Hiding using Digital Images", *Defense Science Journal*, Volume 62, Issue 1, 2012, pp. 11–18.
28. C.R. Geetha, H. D. Giriprakash, "Image Steganography by Variable Embedding and Multiple Edge Detection using Canny Operator", *International Journal of Computer Applications*, Volume 48, Issue 16, 2012, pp. 15–19.
29. Youssef Bassil, "Image Steganography based on a Parameterized Canny Edge Detection Algorithm", *International Journal of Computer Applications*, Volume 60, Issue 4, 2012, pp. 35–40.
30. Vijay Kumar Sharma, Vishal Shrivastva, "A steganography algorithm for hiding image in image improved LSB substitution by minimize Detection", *Journal of Theoretical and Applied Information Technology (JATIT)*, Volume 36, Issue 1, 2012, pp. 1–8.
31. Atallah M. Al-Shatnawi, "A New Method in Image Steganography with Improved Image Quality", *Applied Mathematical Sciences*, Volume 6, Issue 79, 2012, pp. 3907–3915.
32. Rig Das, Themrichon Tuithung, "A novel steganography method for image based on Huffman Encoding", *Emerging Trends and Applications in Computer Science (NCETACS), 3rd National Conference (IEEE)*, March 2012, pp. 14–18.
33. Nitin Jain, Sachin Meshram, Shikha Dubey, "Image Steganography Using LSB and Edge – Detection Technique", *International Journal of Soft Computing and Engineering (IJSCE)*, Volume 2, Issue 3, 2012, pp. 217–222.
34. A. Swathi, S. A. K. Jilani, "Video Steganography by LSB Substitution Using Different Polynomial Equations", *International Journal of Computational Engineering Research*, Volume 2, Issue 5, 2012, pp. 1620–1623.
35. K. Naveen BrahmaTeja, G. L. Madhumati, K. Rama Koteswara Rao, "Data Hiding using EDGE based steganography", *International Journal of Emerging Technology and advanced Engineering*, Volume 2, Issue 11, 2012, pp. 285–290.
36. Wen-Jan Chen, T. Chin-Chen Chang Hoang Ngan Le, "High payload steganography mechanism using hybrid edge detector", *Expert Systems with Applications (Elsevier)*, Volume 37, Issue 4, 2010, pp. 2702–3528.
37. A. P. Sherly and P. P. Amritha, "A Compressed Video Steganography using TPVD", *International Journal of Database Management Systems (IJDMS)*, Volume 2, Issue 3, 2010, pp. 67–80.
38. Yeuan-Kuen Lee, Graeme Bell, Shih-Yu Huang, Ran-Zan Wang, and Shyong-Jian Shyu, *An Advanced Least-Significant-Bit Embedding Scheme for Steganographic Encoding*, Springer, 2009, pp. 349–360.

Chapter 13

Machine learning techniques for IoT applications in healthcare

Intelligent data-drive flow

Anupam Bonkra, Pummy Dhiman and Jai Sonker

13.1 INTRODUCTION

The pro-active integration of Machine Learning (ML) and the Internet of Things (IoT) is bringing change to the healthcare segment through new concepts that are expected to dramatically alter patient care, diagnosis, and treatment plans. IoT enabled devices which are in a position to compile as well as upload latest health information in combination with the computing power of ML holds priceless opportunity to revolutionize healthcare services. Health-care solutions/ ML applications when implemented with IoT technology have become one of the potential forces to transform this sector [1]. Using IoT in healthcare is called healthcare Internet of things or H-IoT. This synergistic convergence enables the use of H-IoT devices such as wearable, sensors, and smart medical equipment to gather massive amounts of real-time data. On the other hand, the ML algorithms process this raw data with the intention of gaining useful pattern, knowledge, and forecast, all ingredients that will revolutionize healthcare through customization of services and anticipatory healthcare. In the recent past, the use of gadgets such as wearable devices, smart sensors lastly remote monitoring systems have led to great expansion of data access in the health industry [2]. These H-IoT devices produce a continuous flow of data, including several aspects such as vital signs, behaviors, and even environment. The use of Machine Learning techniques therefore 'applies brute intelligence' in tackling this huge and complex data that can be used as the basis for making informed decisions, arriving at probable trends in health and other related fields. Starting from the use of ML for disease forecasting and ending with precise treatment plans – the combination of technologies creates the wide range of opportunities for active and targeted patient approach [3].

As a single combination, ML and IoT bring benefits that consist of new opportunities; however, they also pose certain problems like data security, data interoperability, and data ethics. Topicality of using data analytics for

DOI: 10.1201/9781003590439-13

patients' benefit, while at the same time preventing potential compromises of their rights and privacy, emerges as another major challenge in this emergent environment [4]. This paper aims to explore and discuss the IoT and ML for healthcare by developing and providing more insights and ideas about it to the current study to contribute to the discussion in this area. Real-world implementations, technological synergies, and the evolving landscape of healthcare informatics will be explored, shedding light on the transformative power of these intelligent technologies. The discussion encompasses not only the micro-level impact on individual patient outcomes but also the macro-level optimization of healthcare systems, resource allocation, and operational efficiency [5].

13.1.1 Architecture of healthcare Internet of Things (H-IoT)

Healthcare Internet of Things (H-IoT) is the term used sometimes to describe the Internet of Things (IoT) branch dedicated to medical research. A healthcare Internet of things system is an example of a network of all available health resources that are connected to one another to enable the rapid flow of data between them over the Internet [6]. This means that all medical resources—doctor offices, hospitals, rehabilitation facilities, all medical equipment and sensors, and patients—will integrate with each other to enable real-time data transmission while preserving an ongoing information flow. The multitude of sensors, in conjunction with signal interpretation software, may detect anomalies and transmit patient information to physicians and healthcare facilities for diagnostic and analytical reasons. Remedial action may then be suggested and implemented [7]. The H-IoT architecture was described in Figure 13.1.

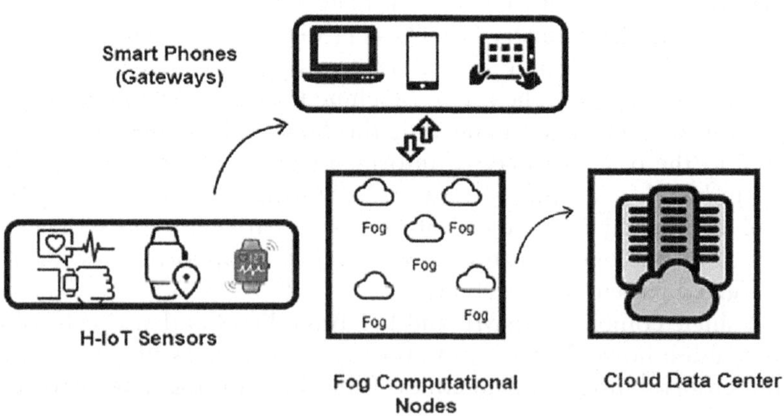

Figure 13.1 Architecture of H-IoT.

An end-to-end network with three main operating layers commonly makes up an H-IoT system:

- IoT sensor layer: The Internet of Things (IoT) sensor layer is comprised of real-world devices such as sensors, actuators, and smart devices. These devices are responsible for monitoring and gathering data, which is then sent to the transport layer [8].
- Transport layer: This is the layer that deals with the secure transfer of data and this move from the sensor part of the IoT layer to the processor either through wired or wireless networks. Serving the main Internet of Things sensor layer is the transport layer as the environment which is open to every person. To this end, there are numerous forms of communication technology that may be used, which include the Wi-Fi, Bluetooth, Zigbee and so on. For the medium-range communication technology like Wi-Fi is adopted in this investigation. Internet Protocol version 4 (IPv4) was adopted as the transport layer protocol while IEEE 802 was used as the ETHER TYPE [9]. 11 (WLAN) was used as Internet of Things sensor protocol; Also, wireless local area networks (WLAN) are employed to enable the forwarding of data from the edge unit to the on-demand layer.
- Edge layer: This layer boosts the Internet of Things systems in terms of efficiency of operation, security, and expansion. For the purposes of analysis and illness forecasting, it consists of gateway devices that are responsible for acquiring information from other smart devices of the IoT that are included in the infrastructure of the network or from other original sources that are delivered by users [10]. This information is further passed on to the on-demand layer.
- On-demand layer: This layer—also referred to as the cloud layer—houses the server that analyzes data, applies advanced analytics to it, and sends back data from the Internet of Things sensor layer across the transport layer, along with additional user-supplied data. This cloud layer's application is housed on a Kubernetes cluster.

This paper provides a thorough examination of the intersection between machine learning techniques and H-IoT applications in the healthcare sector, offering insights into their collaborative potential. The following are the contribution of this study.

- Emphasizes the need for integrating ML algorithms as part of the data collection and analysis to enable real-time data monitoring through IoT devices and accurate disease diagnosis and treatment planning.
- Showcases the application of ML in combination with IoT to improve patient's lives and enhance treatment planning and preventative measures for chronic conditions by constantly monitoring and analyzing data on parameters.

- Demonstrates the synchronization of ML and IoT which transforms the health care sector by reallocation of resources and efficient use of health care services, hence leading to the enhancement of the health care system.
- Explains in detail how machine learning methods can be used to predict the development of a certain disease and the ways of its prevention and treatment to provide better care for the patient.
- Admits the difficulties of connecting ML, IoT and recognizes data security issues, problems of compatibility, the problem of data ethics, which should be solved to protect patient's rights and his/her personal data.

The rest of this work is organized as follows. In Section 13.2, related work is highlighted. Section 13.3, sheds light on ML algorithms used to process data in IoTs and in Section 13.4 working of ML is explained. Section 13.5 comprises various IoT technologies used in healthcare and Section 13.6 sheds light on the use cases. Section 13.7 concludes this study.

13.2 LITERATURE REVIEW

This research focuses on the implementation of deep learning applications while examining many technical strategies to enhance the Internet of Things (IoT) in the healthcare sector. In order to create medical treatments, this project aims to investigate the many aspects of deep learning-based Internet of Things (IoT) healthcare technologies and provide a range of deep learning-based frameworks, architectures, platforms, and algorithms. The paper also provides a thorough analysis of research endeavors focused on the potential contributions of deep learning to machine health monitoring, telehealth and ambient assisted living systems, human activity detection, vital sign collection for patients, and data fusion activities [11]. This chapter discusses how the internet of things and machine learning might completely change how traditional healthcare is provided. Several varieties of machine learning techniques are discussed, such as unsupervised learning, semi-supervised learning, reinforcement learning, and unsupervised learning. Furthermore, the Internet of Things' layered architecture—which is specifically designed for healthcare services—is also covered. Numerous applications, such as smart diagnostic care, personal care, and the treatment of chronic illnesses, have been discussed in connection to the integration of machine learning with the internet of things. An IoT- and machine learning-based solution is a feasible choice for the implementation of dependable, secure, and dispersed healthcare applications. Furthermore, blockchain technology will allow for the integration of machine learning (ML) and the Internet of Things (IoT). A transparent data flow service, an immutable data record, safe data transfers between persons and organizations, secured transactions, and a transaction validation process are just a few advantages offered by blockchain technology. One of the features of Blockchain technology is the creation of a smart contract, which may be used to form a real agreement in the healthcare industry. Blockchain, machine learning, and Internet of Things use in the healthcare industry will focus on claim management, pharma supply chain management, and smart health supply chain management. [12, 13]. The most crucial IoT and AI apps and technologies that are recommended to be used to build a safer, more accurate, and more predictive healthcare system have been covered throughout this review. It has been observed that the designs under examination generally show a propensity to shift information processing closer to the edge. This might be because there are more resources than ever accessible for edge and fog devices. Even if most main inferencing is currently done in the cloud, with the release of upcoming libraries like TinyML and specialized hardware like Tensor processing units, this is expected to shift more and more towards the edge. Even though a number of creative ways are being put forward, some solutions are not practical to apply. This is because the healthcare sector demands a high degree of precision and cannot afford to have false negatives. Additionally, patients' trust in the apps and willingness to use them are impacted by their concerns around the security and privacy of their information. Despite this, it is expected that in the not too distant future, patients will have more trust in and understanding about these apps due to the concepts and execution of applications that provide a high degree of data privacy and security [14]. The most crucial IoT and AI apps and technologies that are recommended to be used to build a safer, more accurate, and more predictive healthcare system have been covered throughout this review. It has been observed that the designs under examination generally show a propensity to shift information processing closer to the edge. This might be because there are more resources than ever accessible for edge and fog devices. Even if most main inferencing is currently done in the cloud, with the release of upcoming libraries like TinyML and specialized hardware like Tensor processing units, this is expected to shift more and more towards the edge. Even though a number of creative ways are being put forward, some solutions are not practical to apply. This is because the healthcare sector demands a high degree of precision and cannot afford to have false negatives. Additionally, patients' trust in the apps and willingness to use them are impacted by their concerns around the security and privacy of their information. Despite this, it is expected that in the not too distant future, patients will have more trust in and understanding about these apps due to the concepts and execution of applications that provide a high degree of data privacy and security [15].

They provide an overview of the conceptualization and architectural design work they have done to create the Wearable Internet of Things (WIoT). The architecture of the wearable Internet of Things is divided into three stages: Wearable body area sensors are linked to the human body in order to gather data. 2) With access to the Internet The information collected from the sensor may be moved to the server or cloud via gateways so that it can be stored and analyzed. 3) Support for cloud and big data, which enables machine learning and massive information processing features as well as easier management over wearable data [16]. They propose three layers for the system, which are as follows: a) They provide unprocessed data from medical Internet of Things devices that have a range of sensors built in, such as skin temperature sensors, electrocardiogram (ECG) sensors, and electroencephalogram (EEG) sensors. Data categorization, information filtering, and processing all happen at this level. Additionally, data analysis and prediction are provided at this level [17]. The writers of the paper provide an overview of the Internet of Things (IoT) and its development, both now and in the future. The Internet of Things (IoT) has developed from its inception as Electronic Data Interchange (EDI) in 1999 to the Internet of People (Internet, M-internet), and it is now referred to as the Internet of Things. In addition to improving many different industries, IoT also gives our lives more intelligence [18]. They demonstrate the services provided by Internet of Things-based medical equipment that are linked to patients, doctors, and the internet in the healthcare industry. The human body is connected to these gadgets. There is a claim that a model that takes a thorough approach to the Internet of Things (IoT) in e-health might provide the most intelligent medical care available. The proposed idea, which is based on the Internet of Things, consists of four steps: This step is important because it makes it easier to get in touch with medical facilities, which in turn improves healthcare services for the rural population. A) Patient Records that contain all patient data, including datasets and real-time data; B) Clinical Decision Systems that give doctors DSS based on linked knowledge; C) Remote patient monitoring using sensors attached to the patient's body to gather data; and D) Remote patient treatment and monitoring [19]. The writers provide the idea of how the Internet of Things (IoT) enhances the environment for healthcare. The three layers that make up the suggested system are as follows: Real-time data collection and information collecting is the responsibility of the first layer, sensing. Numerous sensors, including temperature and pulse rate sensors, are employed to do this. This amount of data transmission from sensors to a data server enables the information to be assessed before being converted into a chart analysis. The Tier-2 sending is what is meant by this. The doctor may access the data server and diagnose various conditions using Tier-3 accessing. An ATMEL 89s52 microcontroller, an analog-to-digital converter, an IC-7805 voltage sensor, a temperature sensor, a pulse rate sensor, RS232, and other hardware components are included in the proposed work. [20]. A web page application and an android application are also included in the software components. They provide a healthcare monitoring system that uses classifier algorithms and the internet of things (IoT) to generate predictions. For older people, this approach offers great self-monitoring, especially for those with chronic diseases including cancer, diabetes, stroke, and other ailments of a similar kind.

The suggested strategy for keeping an eye out for stroke victims is divided into three levels: The microprocessor, blood pressure sensor, and sugar sensor are three of the parts that comprise the hardware layer. 3) The application layer (a web environment or a cloud server), as well as machine learning techniques like Naive Bayes and Random Forest. The Random Forest approach is implemented, and the result is an accuracy of 93% [21]. The various approaches and contributions of ML and IoT in healthcare are compared in the following table (Table 13.1).

Table 13.1 Comparison of different approaches of deep learning in healthcare

Study	*Focus*	*Methodologies*	*Key contributions*
Bolhasani et al.	Deep Learning in IoT for Healthcare	Deep learning frameworks, architectures, platforms	Comprehensive review of deep learning applications in telehealth, ambient assisted living, machine health monitoring, and data fusion.
Agarwal et al.	ML and IoT in Healthcare	Reinforcement learning, semi-supervised learning, unsupervised learning, blockchain integration	Discusses integration of ML with IoT, blockchain for secure transactions, and applications in smart diagnostics and chronic disease treatment.
Bharadwaj et al.	AI and IoT in Healthcare	Edge computing, cloud-based processing, TinyML, Tensor Processing Units	Emphasizes edge processing, high accuracy needs, data privacy and security, and future developments in healthcare apps.
Pandey	Smart Health Monitoring System	IoT, ML techniques, Android and web applications	Proposes a system for smart health monitoring using IoT and machine learning, highlights patient privacy and security concerns.
Hiremath et al.	Wearable IoT Architecture	Data collection, gateway connectivity, cloud support	Details the architecture of wearable IoT for healthcare, including data collection, storage, and machine learning support.
Darshan et al.	IoT in Health Sector	Raw data processing, data classification, analysis	Three-tier architecture for IoT in healthcare, including raw data collection and analysis.
Maqbool et al.	Evolution of IoT	IoT in healthcare evolution, current and future systems	Overview of IoT's evolution and its impact on medical treatment and healthcare devices.

(Continued)

Table 13.1 *(continued)*

Study	*Focus*	*Methodologies*	*Key contributions*
Chatterjee et al.	Comprehensive IoT Approach in E-health	Patient records, clinical decision support, remote monitoring	Four-step model for IoT in E-health focusing on patient records, decision support, and remote monitoring.
Budida et al.	IoT-based Healthcare System	Real-time data collection, data server analysis, physician access	Three-level system for real-time data collection and analysis, hardware and software components described.
Kumar	mHealth and 6LoWPAN-based Healthcare	Data acquisition, storage, processing, 6LoWPAN	Describes mHealth framework and 6LoWPAN-based systems for healthcare data management.
Ani et al.	IoT and Classifier Algorithms for Prediction	IoT monitoring, Naive Bayes, Random Forest	Monitoring system for chronic illnesses with a high accuracy rate using Random Forest.
Onasanya et al.	IoT-based Cancer Care System	Wireless sensor network, Zigbee technique, cloud storage	Proposes a cancer care system using IoT and Zigbee for data analysis and cloud storage.
Dey et al.	Security in IoT Healthcare Systems	Security challenges, solutions	Analyzes security challenges in IoT healthcare systems and proposes solutions.
AlMansour et al.	IoT in Saudi Arabia Healthcare	IoT framework for health sector	Proposes a framework for IoT-based healthcare in Saudi Arabia, based on literature review.
Ed-daoudy et al.	Cardiac Disease Prediction System	Apache Spark for data processing, machine learning	Uses Apache Spark for processing and predicting cardiac disease from various data sources.

13.3 ALGORITHMS OF MACHINE AND DEEP LEARNING

Decision-makers at the data processing layer typically apply methods rooted in artificial intelligence to examine data obtained from the latter layer. They may be further classified as supervised learning algorithms as the input dataset contains the classification labels or as the unsupervised learning algorithms. In general, the machine learning algorithms are usually more focused on labeled data while deep learning algorithms are more efficient when it comes to using the unlabeled data.

13.3.1 Multi-layer perceptron (MLP)

Multilayer perceptrons, often known as MLPs, are a form of feed-forward artificial neural network (ANN) in which data is processed in a sequential manner, beginning with input and ending with output. The input layer, the hidden layers, and the output layer are the three layers that make up an MLP — all of which are completely linked to one another. The principal output layer is designed to produce the results of the conclusion phase including prediction while the input layer enables the network to receive the initial signal that needs to be processed. MLP's hidden layers are distinguishable, and their number can differ; however, they are situated between input and output layers, and the actual complexity of calculations is concentrated there. All these layers are hidden from the user and, therefore, he is only able to interact with the GUI layer. In the case of multilayer perceptron (MLP) networks, various types of activation or transfer functions are incorporated with the specific aim of determining the output of the network. Some of these functions include; ReLU (Rectified Linear Unit), Tanh, Sigmoid & Softmax. For this purpose, one of the techniques of supervised learning such as backpropagation is employed with the aim of training the neurons included in the create neural network's MLP layer. Indeed, the usage of MLPs is rather beneficial for applying to problems that have basic regression, pattern classification, identification, prediction, and approximation problems [22].

13.3.2 Convolutional neural network

Popularly referred to as ConvNet or CNN, Convolutional neural networks is a type of neural network model designed for handling two-dimensional image data. However, it is also fit for dealing with one dimensional as well as three dimensions data as well. The convolution process originally is a math linear operation that takes two functions to produce a third function which describes how the form of one function is altered by the other and this is where the CNN gets its name from. CNN also referred to as Convolutional neural networks. In even more plain English, convolution can be described as how it is possible to get details concerning a picture using two matrices which are images. This technique is referred to as feature extraction and it is applied in order to find out various characteristics of a picture or at least such parts of a picture as corners or edges. Next, the feature map is forwarded to the subsequent layers in order that these layers may gain new features of the enter picture [23].

The pooling layer is laid subsequent to the convolution layer and its goal is to bring the size of the feature map after convolution lowers the computation complexity. Besides, it works independently for each feature map, which can assist in reducing the number of connections between the layers. Of all the pooling procedures, average pooling and maximum pooling are the most commonly applicable ones. Following that, the output in pooling is again unfolded in the form of a vector and the vector goes through the other processing phases. Neurons that link betwixt the several layers are connected through the fully connected (FC) layer, which comprises weights, bias, and neurons. As it can be seen, it is feasible for a neural network to classify and identify pictures if the flattening process result is input to it. Activation functions and dropout functions are two features that are used in CNNs. Sometimes a model fits the training data very well, but very poorly on new data using the same features; this is overfitting. Dropout is a technique that resolves this problem. The dropout layer is used in order to decrease the complexity of the chosen architecture and minimize overfitting. It accomplishes this through dropout which is the procedure of letting some neurons 'die' at random during the entire training phase. That is, a dropout rate of 0. 2 means that randomly 20% of the nodes are removed or they are set to become inactive [24]. These activation functions are ReLU, Softmax, tanH, and Sigmoid functions and these decide which information is to be transmitted within the network. When it comes to binary classification, Sigmoid and Softmax are some of the suggested methods, whereas Softmax is often the favored method for multi-class classification. Computer networks are particularly effective at performing computer vision tasks such as image analysis, identification of images and videos, categorization of images, and natural language processing.

13.3.3 Recurrent neural network (RNN)

The distinctive feature of recurrent neural networks (RNNs) is that they use the output from the preceding step as the input for the next phase. RNNs are quite accurate in predicting following data points because of their internal memory, which enables them to retain significant elements of the input. Because of this feature, RNNs are the recommended approach for sequential data—that is, ordered data where related things appear one after the other. Examples of sequential data include time series, financial data, voice, text, audio, video, and weather patterns. An RNN uses a looping information cycle to enable decision-making based on both the current input and the knowledge gained from earlier inputs that are looped back into the network. RNNs are widely utilized in daily applications; they form the basis of programs such as Google Translate and Siri. Although each of the three models—MLPs, CNNs, and RNNs—has advantages, no one model is ideal for every job. For this reason, a lot of deep learning models use a mix of MLPs, CNNs, and RNNs to tackle certain tasks [25].

13.3.4 Logistic regression

Logistic regression is a kind of method belonging to statistical analysis and is commonly applied to classification problems with binary values.

The objective of this technique is to predict the likelihood that an observation comes from one of the two classes possible. Sigmoid or logistic function is used in the construction of the relationship between the dependent variable, as well as the independent variables during the application of the logistic regression model. The sigmoid function is used for probability modeling because owing to its property which states that, the sigmoid of any real-valued integer maps it to the range between 0 and 1. A function which maps any integer between 0 and any real number to the interval [0,1] which is useful mostly for probabilities [26].

13.3.5 Decision tree

Decision tree is of the data model type, which has characteristics similar to that of trees where each node is a characteristic or attribute, the branches whose path forms rules and the end nodes or leave nodes that represent results or class labels. The decision tree method locates the region of the feature space in a step-wise manner, and it aims to minimize imperfection at each stage or to maximize the amount of acquired information. As a result, we presume that X stands for the feature space while Y stands for the target variable [27]. The method chooses a feature x i and a threshold θ at each node t in order to divide X into two subsets: He left side and right side. These subsets are termed as X left and X right. This procedure will go on until the tree attains one of the stopping criteria, for instance, the tree reaches the maximum depth or the minimum cases per node. This process aims at achieving the condition of making SSs as homogeneous as possible concerning feature space and target variable Y. The splitting is determined by:

$$X_{left}(t) = \{x \in X \mid x_i \leq \theta\}$$

$$X_{right}(t) = \{x \in X \mid x_i > \theta\}$$

This process of splitting continues in a recursive manner up to the time when a certain condition of stopping will be reached; this could be in terms of a pre- stipulated maximum tree depth or minimum sample size. The use of this algorithm entails browsing the decision tree from the roots down to the leaves in order to provide forecasts. It does this by being conformal to the decision rules that are specified by the features of the input data. Determination of the result to be predicted follows the majority class (classification models) or average or mean of training cases in the node (regression models) [27]. It is also easily understandable, computational simplicity, and ability to handle nominal and numerical data make it popular among the business organization. However, they can be vulnerable to overlearning and these aspects can be gotten rid of by the use of techniques such as pruning, ensemble methods and proper termination strategies.

13.3.6 Natural language processing

One of the branches of artificial intelligence that uses the term natural language processing is aimed at making computers to work with texts like human beings do. Communication can be by its very nature unorganized, may be in form of e-mail, blogs, social media posts, text messages, and many others. However, as indicated above, this unstructured type of communication is quite the opposite and cannot be understood by the machines. The processing of this kind of information and conversion into a text format is where NLP comes in handy as computers can grasp it, thus applying it in tasks such as text creation, analysis, and categorization. These procedures must be applied on the input text data; however, pre-processing of text data is required before these are applied. Depending on the type of jobs that are being done, the specific steps that are involved in text pre-processing can slightly differ, though, there are normally three main steps; text pre-processing, feature extraction and modelling [28]. Text preprocessing entails conversion of the text to lowercase, and the removal of any special characters, punctuation and stop words. To obtain features from the processed text, techniques such as tokenization of the text, stemming, lemmatization, and vectorization of the written language are employed as an attempt to transform the information into numerical value that could be interpreted by the machine learning algorithms. Lastly, modeling involves issues such as textual production, sentiment analysis, and classification by applying machine learning algorithms on the collected characteristics.

13.4 WORKING OF ML IN HEALTHCARE

The application of machine learning with connecting IoT devices to gather and analyze timely health data is known as machine learning in healthcare IoT. Examples of collected data include vital signs, activity level, and others gathered by Internet of Things (IoT), wearables and medical monitors. The information sample passed through preprocessing is inserted into machine learning models. These as models that are designed from previous data are capable of predicting or making decisions in real time depending on the field of usage in the healthcare sector in terms of illness diagnosis, treatment recommendations and distribution of health care facilities. Frequency enhances the accuracy of the model and enables healthcare personnel to make sound decisions. As a result, various challenges such as data security and model explanatory ability can only be properly addressed to warrant a secure and efficient implementation. The basic architecture has been elaborated by Figure 13.2

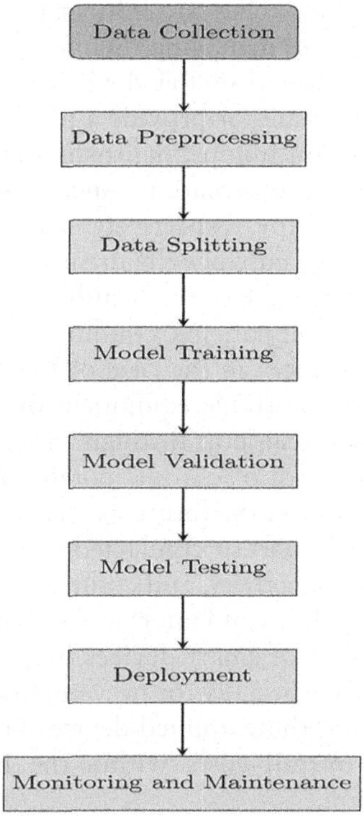

Figure 13.2 Basic architecture of ML in Healthcare.

13.5 DIFFERENT IOT TECHNOLOGIES USED IN HEALTHCARE

In the healthcare domain, IoT technologies are touching the ways the medical data is gathered, correlated, and used for the advancement of patient care, increased health care organization productivity, and for the telemonitoring. Here are some of the key IoT technologies used in healthcare: Here are some of the key IoT technologies used in healthcare.

Wearable Devices: These are arguably the most characteristic IoT devices with regards to the healthcare sector. Smartwatches and other wearables include fitness trackers, and medical wearables that can track signs such as heart rate, blood pressure, activity, sleep, and much more. With these, patient's health can be checked repeatedly, abnormalities can be detected early and health needs can be prevented [29].

Medical Sensors: Medical sensors in the context of IoT are applied to track the different parameters of the human body and the surrounding environment in real time. These sensors can therefore be wired into the medical equipment, the equipment in the hospital, or the rooms where the patient is admitted to, monitor items such as temperature, humidity, oxygen, blood glucose levels, and ECG [30]. They are informative in diagnosing the injuries, treating the patients, and managing them.

Implantable Devices: In this case, IoT devices are placed within the body to diagnose certain health issues or provide a particular treatment [31]. For instance, pacemakers, insulin pumps, neurostimulators, and cardiac monitors with IoT sensors collect information concerning the state of health and the adjustments of the therapy respectively. They apply the principles of individual medicine and also enhance the dreams of patient satisfaction.

Remote Monitoring Systems: Remote health management systems inform healthcare professionals about patients' condition and their compliance with a prescribed treatment protocol in the case of IoT-based technologies [32]. Such systems often include portable equipment or positive devices that are linked to a central overseeing support through wireless communication. They offer strategies for the control of chronic conditions, and response to any complications that may arise in the future as well as patients' involvement.

Smart Medical Devices: Smart or connected medical devices are generally equipped with sensors, connectivity, and computing capability for performing sophisticated functionalities and interfacing with other devices or system [33]. Some of the examples of smart devices include smart inhalers, smart pill bottles and smart beds especially for preventive action of pressure ulcers. It is important to note that these applied devices facilitate medicine administration, decrease the overall costs within the healthcare systems, and increase patient protection.

Telehealth and Telemedicine Solutions: Telehealth & telemedicine services are the applications of IoT technologies for remote consultations, virtual visits, and remote patient monitoring [34]. Smart appliances in the healing sector, video calling solutions, and mobile health applications give healthcare givers a way through which they can attend to patients irrespective of the location they are located in helping to enhance the delivery of adequate health care to everyone regardless of their geographical location.

Hospital Infrastructure Management: IoT technologies are applied in optimisation of the infrastructural support of a hospital and the improvement of the general functioning of the environment [35]. Applying IoT to track assets and inventory, manage facilities and energy contribute to optimizing hospitals' performance and reducing operating expenses. Some of them include IoT-based asset tracking system for the tracking medical equipments and IoT based building automation system for energy management.

13.6 USE CASES OF IOT AND ML IN HEALTHCARE

IoT systems are predicted to propel major technical developments and are constantly changing. Over time, estimations will grow more accurate as new information, research methodologies, and sophisticated procedures become accessible. These procedures may be further enhanced by contemporary wearable technologies, which will help create a more complex healthcare IoT system that facilitates better decision-making. Making a diagnosis usually entails determining the kind of disease a patient has by looking at symptoms that sensors have recorded.

Reference	*Use case*	*Algorithm*	*Key point*
[36]	Heart disease diagnose	KNN	Real-time diagnosis using wearable IoT
[37]	Lung cancer diagnose	SVM	Hyper-parameter tuning results more accurate method
[38]	Type-2 Diabetes diagnose	Ensemble approach (SVM + Random Forest)	It improves reliability
[39]	Diagnosis Chatbots	SVM, KNN	Easy and budget friendly diagnosis without medical professionals
[40]	IoT based cloud framework for Ebola spread control	Decision Tree	Easy large scale implementation using H-IoT
[41]	Smart electronic health record	Deep learning and NLP	Real-time data collection and decision making based on data collection through smart wearables

13.7 CONCLUSION

Machine learning (ML) and the Internet of Things (IoT) have been brought together in the healthcare industry, which has opened up a wide range of opportunities for bringing about a revolution in patient care and healthcare administration. Internet of Things devices are able to not only gather large quantities of data but also extract actionable insights, which may lead to preventative interventions and individualized therapies. This is made possible by the integration of the following machine learning algorithms. Some of the application of machine learning in the internet of things (IoT) in healthcare includes; several use cases that have been hitherto addressed include monitoring of patients remotely and treatment of chronic diseases. However, there is a long way to go before this efficient medium can become a mass medium popular with a vast population. Such challenges are for instance; data security, integration of the models, and interpretability of the models. Closely related to the idea of uncertainty, it is crucial to emphasize those ethical issues, data protection, and compliance with applicable legislation should remain a priority despite the organization's further growth through combining machine learning and the internet of things. However, if we wish to ensure that the improvement of human and socio-fabric health does not become a technologically captured concept and that these technologies are properly deployed and exploited in healthcare ecosystems, we need to priorities the above things.

REFERENCES

1. Bonkra, A., & Dhiman, P. (2021), December. IoT security challenges in cloud environment. In *2021 2nd international conference on computational methods in science & technology (ICCMST)* (pp. 30–34). IEEE.
2. Ahmadi, H., Arji, G., Shahmoradi, L., Safdari, R., Nilashi, M., & Alizadeh, M. (2019). The application of internet of things in healthcare: A systematic literature review and classification. *Universal Access in the Information Society*, *18*, 837–869.
3. Kumar, M., Kumar, A., Verma, S., Bhattacharya, P., Ghimire, D., Kim, S. H., & Hosen, A. S. (2023). Healthcare Internet of Things (H-IoT): Current trends, future prospects, applications, challenges, and security issues. *Electronics*, *12*(9), 2050.
4. Zikria, Y. B., Afzal, M. K., Kim, S. W., Marin, A., & Guizani, M. (2020). Deep learning for intelligent IoT: Opportunities, challenges and solutions. *Computer Communications*, *164*, 50–53.
5. Dang, V. A., Vu Khanh, Q., Nguyen, V. H., Nguyen, T., & Nguyen, D. C. (2023). Intelligent healthcare: Integration of emerging technologies and Internet of Things for humanity. *Sensors*, *23*(9), 4200.
6. Khanna, A., & Kaur, S. (2020). Internet of things (IoT), applications and challenges: a comprehensive review. *Wireless Personal Communications*, *114*, 1687–1762.
7. Ballard, Z., Brown, C., Madni, A. M., & Ozcan, A. (2021). Machine learning and computation-enabled intelligent sensor design. *Nature Machine Intelligence*, *3*(7), 556–565.
8. Kumar, N. M., & Mallick, P. K. (2018). The Internet of Things: Insights into the building blocks, component interactions, and architecture layers. *Procedia Computer Science*, *132*, 109–117.
9. Gupta, B. B., & Quamara, M. (2020). An overview of Internet of Things (IoT): Architectural aspects, challenges, and protocols. *Concurrency and Computation: Practice and Experience*, *32*(21), e4946.
10. Chen, B., Wan, J., Celesti, A., Li, D., Abbas, H., & Zhang, Q. (2018). Edge computing in IoT-based manufacturing. *IEEE Communications Magazine*, *56*(9), 103–109.
11. Bolhasani, H., Mohseni, M., & Rahmani, A. M. (2021). Deep learning applications for IoT in health care: A systematic review. *Informatics in Medicine Unlocked*, *23*, 100550.
12. Agarwal, N., Singh, P., Singh, N., Singh, K. K., & Jain, R. (2021). Machine learning applications for IoT healthcare. *Machine Learning Approaches for Convergence of IoT and Blockchain*, *11*, 129–144.
13. Singh, K. K., Singh, A., & Sharma, S. K. (Eds.). (2021). *Machine learning approaches for convergence of IoT and blockchain*. John Wiley & Sons.

14. Bharadwaj, H. K., Agarwal, A., Chamola, V., Lakkaniga, N. R., Hassija, V., Guizani, M., & Sikdar, B. (2021). A review on the role of machine learning in enabling IoT based healthcare applications. *IEEE Access*, *9*, 38859–38890.
15. Pandey, H., & Prabha, S. (2020, February). Smart health monitoring system using IOT and machine learning techniques. In *2020 sixth international conference on bio signals, images, and instrumentation (ICBSII)* (pp. 1–4). IEEE.
16. Hiremath, S., Yang, G., & Mankodiya, K. (2014, November). Wearable Internet of Things: Concept, architectural components and promises for person-centered healthcare. In *2014 4th international conference on wireless mobile communication and healthcare-transforming healthcare through innovations in mobile and wireless technologies (MOBIHEALTH)* (pp. 304–307). IEEE.
17. Osama, M., Ateya, A. A., Sayed, M. S., Hammad, M., Pławiak, P., Abd El-Latif, A. A., & Elsayed, R. A. (2023). Internet of medical things and healthcare 4.0: Trends, requirements, challenges, and research directions. *Sensors*, *23*(17), 7435.
18. Tiwari, V. K., & Singh, V. (2016). Study of Internet of Things (IoT): A vision, architectural elements, and future directions. *International Journal of Advanced Research in Computer Science*, *7*(7), 22–31.
19. Udoh, I. S., & Kotonya, G. (2018). Developing IoT applications: challenges and frameworks. *IET Cyber-Physical Systems: Theory & Applications*, *3*(2), 65–72.
20. Budida, D. A. M., & Mangrulkar, R. S. (2017, March). Design and implementation of smart HealthCare system using IoT. In *2017 international conference on innovations in information, embedded and communication systems (ICIIECS)* (pp. 1–7). IEEE.
21. Alshammari, H. H. (2023). The internet of things healthcare monitoring system based on MQTT protocol. *Alexandria Engineering Journal*, *69*, 275–287.
22. Taud, H., & Mas, J. F. (2018). Multilayer perceptron (MLP). *Geomatic approaches for modeling land change scenarios* (pp. 451–455), Wiley.
23. Ketkar, N., Moolayil, J., Ketkar, N., & Moolayil, J. (2021). Convolutional neural networks. *Deep learning with Python: Learn best practices of deep learning models with PyTorch* (pp. 197–242), IEEE.
24. Dubey, S. R., Singh, S. K., & Chaudhuri, B. B. (2022). Activation functions in deep learning: A comprehensive survey and benchmark. *Neurocomputing*, *503*, 92–108.
25. Dhruv, P., & Naskar, S. (2020). Image classification using convolutional neural network (CNN) and recurrent neural network (RNN): A review. In *Machine learning and information processing: proceedings of ICMLIP 2019* (pp. 367–381).
26. Nick, T. G., & Campbell, K. M. (2007). Logistic regression. *Topics in biostatistics* (pp. 273–301).
27. Song, Y. Y., & Ying, L. U. (2015). Decision tree methods: Applications for classification and prediction. *Shanghai Archives of Psychiatry*, *27*(2), 130.
28. Nadkarni, P. M., Ohno-Machado, L., & Chapman, W. W. (2011). Natural language processing: An introduction. *Journal of the American Medical Informatics Association*, *18*(5), 544–551.

29. Bayoumy, K., Gaber, M., Elshafeey, A., Mhaimeed, O., Dineen, E. H., Marvel, F. A., ... & Elshazly, M. B. (2021). Smart wearable devices in cardiovascular care: Where we are and how to move forward. *Nature Reviews Cardiology*, *18*(8), 581–599.
30. Wan, J., AAH Al-awlaqi, M., Li, M., O'Grady, M., Gu, X., Wang, J., & Cao, N. (2018). Wearable IoT enabled real-time health monitoring system. *EURASIP Journal on Wireless Communications and Networking*, *2018*(1), 1–10.
31. Darwish, A., & Hassanien, A. E. (2011). Wearable and implantable wireless sensor network solutions for healthcare monitoring. *Sensors*, *11*(6), 5561–5595.
32. Mohammed, M. N., Desyansah, S. F., Al-Zubaidi, S., & Yusuf, E. (2020, February). An internet of things-based smart homes and healthcare monitoring and management system. In *Journal of physics: conference series* (Vol. 1450, No. 1, p. 012079). IOP Publishing.
33. Pramanik, P. K. D., Upadhyaya, B. K., Pal, S., & Pal, T. (2019). Internet of things, smart sensors, and pervasive systems: Enabling connected and pervasive healthcare. In *Healthcare data analytics and management* (pp. 1–58). Academic Press.
34. Haleem, A., Javaid, M., Singh, R. P., & Suman, R. (2021). Telemedicine for healthcare: Capabilities, features, barriers, and applications. *Sensors international*, *2*, 100117.
35. Uslu, B. Ç., Okay, E., & Dursun, E. (2020). Analysis of factors affecting IoT-based smart hospital design. *Journal of Cloud Computing*, *9*(1), 67.
36. Deekshatulu, B. L., & Chandra, P. (2013). Classification of heart disease using k-nearest neighbor and genetic algorithm. *Procedia technology*, *10*, 85–94.
37. Bonkra, A., Dhiman, P., Kamboj, S., & Kaur, S. (2024). Lung disease diagnosis using machine learning from a bibliometric perspective. In *Convergence of blockchain and internet of things in healthcare* (pp. 1–24). CRC Press.
38. Goyal, P., & Jain, S. (2022, March). Prediction of type-2 diabetes using classification and ensemble method approach. In *2022 International Mobile and Embedded Technology Conference (MECON)* (pp. 658–665). IEEE.
39. Majeed, B. A., Hardan, A. Y., Hardan, B. Y., & Munaf, D. F. (2024). Accurate AI-based chatbot to diagnose heart diseases pre-human doctor consultation. *Revue d'Intelligence Artificielle*, *38*(1).
40. Sareen, S., Sood, S. K., & Gupta, S. K. (2018). IoT-based cloud framework to control Ebola virus outbreak. *Journal of Ambient Intelligence and Humanized Computing*, *9*, 459–476.
41. Harerimana, G., Kim, J. W., Yoo, H., & Jang, B. (2019). Deep learning for electronic health records analytics. *IEEE Access*, *7*, 101245–101259.

Chapter 14

Integration of emerging technologies for digital asset management

G. Boopathi Raja

14.1 INTRODUCTION

The integration of emerging technologies with digital asset management (DAM) presents a significant opportunity to enhance security, efficiency, and transparency in handling digital assets. Within the context of DAM, this chapter examines the beneficial interaction between cutting-edge technologies including blockchain, artificial intelligence (AI), and the Internet of Things (IoT) [1, 2].

Artificial Intelligence (AI) plays a pivotal role in revolutionizing DAM by automating key processes and enabling intelligent asset management. One of the primary applications of AI in DAM is metadata enrichment, where algorithms analyze and tag digital assets with relevant metadata, enhancing searchability and organization [3]. Additionally, AI facilitates content classification, enabling efficient categorization of assets based on their attributes and content, thereby streamlining asset retrieval and management workflows. AI-driven anomaly detection mechanisms further bolster security by identifying suspicious activities and potential threats within the digital asset environment [4].

Blockchain technology emerges as a transformative force in DAM, offering unparalleled levels of security, transparency, and trust [5]. By leveraging blockchain's decentralized and immutable ledger, organizations can establish transparent records of digital asset transactions, ensuring the integrity and authenticity of asset ownership and provenance. Blockchain's tamper-resistant nature makes it particularly suitable for industries where maintaining the integrity of digital assets is critical, such as media, entertainment, and finance [6, 7].

The integration of the Internet of Things (IoT) augments DAM by providing real-time monitoring and data collection capabilities. IoT devices embedded within digital assets or storage systems enable organizations to track asset usage patterns, access history, and performance metrics in real-time [8]. This proactive approach to asset management empowers organizations to identify potential vulnerabilities, optimize resource utilization, and mitigate risks effectively.

DOI: 10.1201/9781003590439-14

By synthesizing these emerging technologies, organizations can adopt a holistic approach to DAM, effectively addressing the multifaceted challenges associated with digital asset security and management [9]. A roadmap for successful integration involves leveraging AI algorithms for metadata enrichment and content classification, implementing blockchain-based systems for transparent asset management, and deploying IoT devices for real-time monitoring and data collection [10].

In an era defined by the exponential growth of digital assets, the management and security of these assets have become paramount concerns for organizations across diverse industries. From multimedia content to sensitive financial data, digital assets form the backbone of modern businesses, necessitating robust systems and methodologies for their effective governance and protection [8].

The development of emerging technologies, including artificial intelligence (AI), blockchain, and the Internet of Things (IoT), offers a significant potential for transforming conventional methods of digital asset management (DAM). This chapter embarks on a comprehensive exploration of the symbiotic relationship between these cutting-edge technologies and DAM, aiming to elucidate their transformative potential and practical implications [9, 10].

At the heart of contemporary DAM lies the challenge of efficiently organizing, securing, and retrieving an ever-expanding array of digital assets. Traditional methods, reliant on manual categorization and centralized storage systems, often struggle to keep pace with the dynamic nature of digital content. As such, there arises a pressing need for innovative solutions capable of addressing the evolving demands of digital asset management.

Artificial intelligence emerges as a cornerstone technology in the quest to streamline DAM processes and enhance operational efficiency. Through advanced algorithms and machine learning techniques, AI facilitates the automated enrichment of metadata, enabling swift and accurate asset categorization. Moreover, AI-driven anomaly detection mechanisms empower organizations to proactively identify and mitigate potential security threats, safeguarding digital assets against malicious activities [3].

In parallel, blockchain technology emerges as a disruptive force, poised to redefine the principles of transparency, security, and trust in digital asset management. By leveraging its decentralized ledger and cryptographic principles, blockchain enables the establishment of immutable records, thereby ensuring the integrity and provenance of digital assets. The decentralized nature of blockchain not only mitigates the risks associated with centralized storage but also introduces novel paradigms for asset ownership and transactional transparency.

Complementing these advancements, the Internet of Things (IoT) revolutionizes DAM by endowing assets with real-time monitoring and data collection capabilities. Embedded sensors and connected devices enable organizations to glean insights into asset usage patterns, access history, and performance metrics, facilitating informed decision-making and proactive asset management strategies [8].

As organizations navigate the intricate landscape of digital asset management, the integration of these emerging technologies emerges as a strategic imperative. By synthesizing the capabilities of AI, blockchain, and IoT, organizations can forge a holistic approach to DAM, fortifying their defenses against emerging threats and optimizing the efficiency of asset management workflows [9].

Through a systematic exploration of the integration of AI, blockchain, and IoT in DAM, this chapter endeavors to equip readers with the insights and tools necessary to navigate the complexities of digital asset management in an increasingly digitized world. By embracing innovation and leveraging the transformative potential of emerging technologies, organizations can embark on a journey towards secure, efficient, and future-ready digital asset management practices [10].

The integration of emerging technologies holds immense potential for transforming the landscape of digital asset management. By embracing AI, blockchain, and IoT solutions, organizations can fortify their digital asset security, optimize operational efficiency, and navigate the evolving landscape of digital asset management successfully [9].

14.2 ARTIFICIAL INTELLIGENCE (AI) IN DIGITAL ASSET MANAGEMENT

Artificial Intelligence (AI) is reshaping the landscape of digital asset management (DAM), offering a suite of capabilities that streamline processes, enhance security, and improve the overall efficiency of managing digital assets [11]. Within the realm of DAM, AI finds application in several key areas, including automated content classification, metadata enrichment, and anomaly detection for threat mitigation [12]. Figure 14.1 show the different stages involved in Digital Asset Lifecycle.

14.2.1 Automated content classification

AI-powered algorithms enable automated content classification, a process crucial for organizing and categorizing digital assets based on their attributes, content, and context. By leveraging machine learning techniques, AI models can analyze the content of digital assets, recognize patterns, and assign relevant tags or labels [13]. This automated classification not only accelerates the categorization process but also ensures consistency and accuracy across a vast array of assets. With AI-driven content classification, organizations can swiftly locate and retrieve assets, optimize content workflows, and facilitate seamless collaboration among teams.

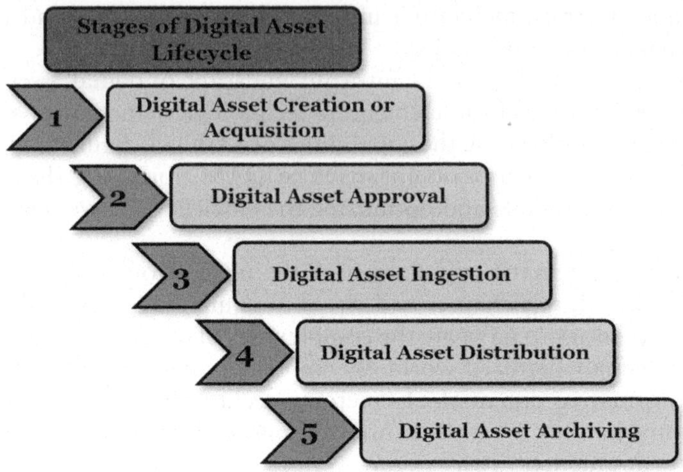

Figure 14.1 Stages of digital asset lifecycle.

14.2.2 Metadata enrichment using AI

Metadata serves as the backbone of effective digital asset management, providing valuable insights into the characteristics, context, and usage rights of assets. AI plays a pivotal role in automating metadata enrichment processes by analyzing asset content, extracting pertinent information, and generating descriptive metadata tags. Through natural language processing (NLP) and computer vision algorithms, AI can identify objects, people, locations, and other relevant attributes within digital assets, enriching metadata to enhance searchability and discoverability. By automating metadata enrichment using AI, organizations can streamline asset management workflows, improve search accuracy, and maximize the utility of digital assets across diverse contexts.

14.2.3 Anomaly detection and threat mitigation

In an increasingly interconnected digital ecosystem, the security of digital assets remains a paramount concern for organizations. AI-driven anomaly detection mechanisms empower DAM systems to proactively identify aberrant behaviors, suspicious activities, and potential security threats within the digital asset environment. By analyzing patterns, deviations, and anomalies in user behavior, access patterns, and asset interactions, AI models can detect unauthorized access attempts, data breaches, and malicious activities in real-time. Through continuous monitoring and analysis, AI-powered anomaly detection strengthens the resilience of DAM systems, enabling organizations to mitigate security risks, safeguard sensitive assets, and uphold data integrity.

AI represents a transformative force in digital asset management, offering unparalleled capabilities to automate content classification, enrich metadata, and detect anomalies for threat mitigation [14]. Through the utilization of AI-powered technologies, organizations have the ability to optimize workflows related to asset management, improve the discoverability and search capabilities, and strengthen the security stance of their digital asset repositories. The incorporation of AI into DAM has the potential to facilitate innovation and uncover novel prospects in the administration of digital assets across various sectors and domains, as AI continues to advance.

14.3 BLOCKCHAIN TECHNOLOGY FOR SECURE DIGITAL ASSET RECORDS

Blockchain technology offers a paradigm shift in digital asset management (DAM) by introducing transparent, tamper-resistant asset records, decentralized management systems, and robust applications in provenance and authenticity verification [15]. Figure 14.2 illustrates the management of Digital Assets for E-commerce. The breakdown of how blockchain technology enhances the security and integrity of digital asset records is listed as follows:

14.3.1 Transparent and tamper-resistant asset records

Blockchain's distributed ledger architecture ensures transparency and immutability in digital asset records. Every transaction or interaction involving a digital asset is securely documented in a block, which is interconnected with preceding blocks to establish a sequential sequence of transactions. By virtue of this transparency, all blockchain network participants are granted access to an identical, synchronized ledger. This eliminates the necessity for intermediaries and substantially mitigates the potential for data manipulation or unauthorized modifications. Blockchain's inherent resistance to tampering guarantees that transactions cannot be retroactively altered without the consent of the majority of participants. This characteristic significantly improves the integrity and reliability of records pertaining to digital assets [16].

14.3.2 Decentralized asset management

Traditionally, digital assets are managed and stored within centralized systems, which are susceptible to single points of failure, cyberattacks, and data breaches. Blockchain technology decentralizes asset management by distributing copies of the ledger across multiple nodes in a network, thereby eliminating the reliance on a single authority or central repository. Decentralized asset management not only enhances resilience and fault tolerance but also democratizes access to digital assets, fostering greater inclusivity and

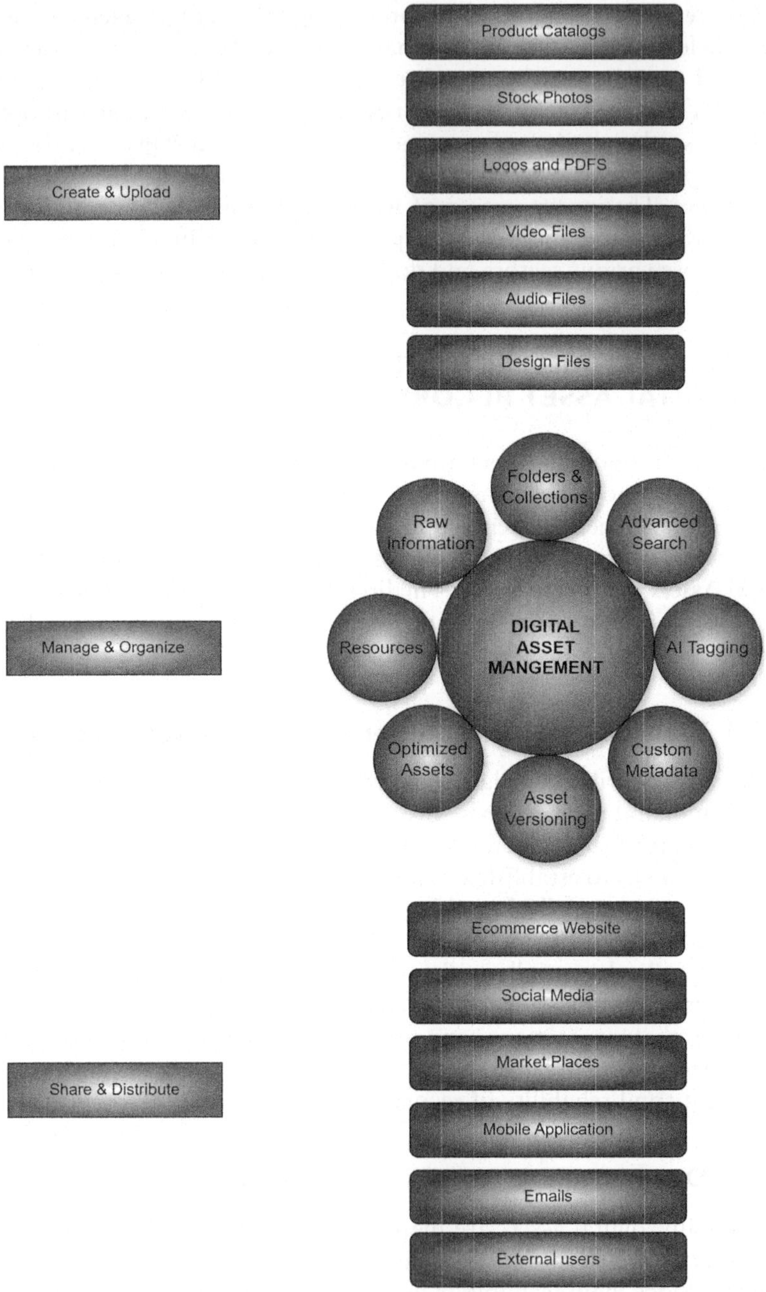

Figure 14.2 Management of digital assets for e-commerce.

accessibility for users across diverse geographical locations and organizational boundaries.

14.3.3 Applications in provenance and authenticity verification

Blockchain technology is particularly well-suited for applications requiring provenance and authenticity verification of digital assets. By recording the entire lifecycle of an asset, including its creation, ownership transfers, and modifications, blockchain enables stakeholders to trace the provenance of digital assets back to their origins with a high degree of certainty. This capability is especially valuable in industries such as luxury goods, art, pharmaceuticals, and supply chain management, where establishing the authenticity and integrity of assets is critical for combating counterfeiting, fraud, and unauthorized duplication. Through cryptographic signatures, smart contracts, and consensus mechanisms, blockchain facilitates secure and transparent verification of asset provenance, enabling stakeholders to make informed decisions and uphold trust in digital transactions [17].

Blockchain technology offers a robust framework for securing digital asset records, decentralizing asset management, and enabling verifiable provenance and authenticity verification. By leveraging blockchain's transparent, tamper-resistant ledger, organizations can enhance the integrity, security, and trustworthiness of their digital asset repositories, unlocking new opportunities for innovation and value creation across diverse industries and domains.

14.4 INTERNET OF THINGS (IOT) FOR REAL-TIME ASSET MONITORING

Digital asset management (DAM) is transformed by the Internet of Things (IoT), which facilitates real-time asset monitoring through the utilization of sensor networks, asset tracking, and data collecting to enable predictive maintenance [18]. In this section, how IoT enhances asset management can be explained.

14.4.1 Asset tracking and location-based services

IoT devices embedded within assets or connected to asset management systems enable precise tracking and monitoring of asset location and movement in real-time. GPS-enabled trackers, RFID tags, and Bluetooth beacons facilitate asset tracking across diverse environments, providing organizations with visibility into the whereabouts of assets throughout their lifecycle.

Location-based services powered by IoT technologies enable asset managers to optimize asset utilization, streamline logistics, and mitigate the risk of loss or theft by monitoring asset movements and implementing geofencing capabilities to define virtual boundaries and trigger alerts when assets deviate from predefined zones.

14.4.2 Sensor networks in DAM

IoT sensor networks play a pivotal role in capturing real-time data on asset performance, condition, and environmental parameters. Various sensors, including temperature sensors, humidity sensors, vibration sensors, and moisture sensors, collect granular data on asset health and operating conditions, enabling proactive maintenance and condition-based monitoring. By deploying IoT sensor networks within digital asset environments, organizations gain insights into asset behavior, identify potential anomalies or deviations from normal operating parameters, and preemptively address issues before they escalate into critical failures or downtime. This predictive approach to asset management enhances operational efficiency, reduces maintenance costs, and prolongs asset lifespan.

14.5 DATA COLLECTION FOR PREDICTIVE MAINTENANCE

The utilization of IoT-enabled data collection enables the implementation of predictive maintenance strategies through the application of machine learning algorithms, historical performance metrics, and real-time sensor data to forecast equipment failures and optimize maintenance schedules [19]. Through the examination of correlations, patterns, and trends within streams of asset data, predictive maintenance models have the capability to detect early warning signs of nearing failures, arrange maintenance operations during the most advantageous periods, and allocate resources in accordance with the criticality and risk profiles of the assets. This proactive maintenance approach minimizes unplanned downtime, maximizes asset availability, and enhances overall reliability, resulting in cost savings and operational efficiencies for organizations across various industries.

The Internet of Things (IoT) facilitates digital asset management (DAM) through the utilization of sensor networks, asset tracking, and data-driven predictive maintenance strategies, all of which enable real-time asset monitoring. Organizations have the ability to optimize asset utilization, improve operational efficiency, and mitigate risks related to asset outage and failure through the utilization of IoT technologies. The ongoing development of IoT and its potential integration into DAM present novel prospects for value generation and innovation, propelling digital transformation in various sectors and domains.

14.6 INTEGRATION STRATEGIES FOR SYNERGISTIC EFFECTS

Integration strategies for achieving synergistic effects between artificial intelligence (AI), blockchain, and the Internet of Things (IoT) in digital asset management (DAM) require careful consideration of the interplay between these technologies, seamless connectivity, and cross-platform compatibility [20, 21]. Figure 14.3 shows different Phases in Digital Asset Management In this section, how organizations can effectively integrate AI, blockchain, and IoT to maximize the benefits of DAM can be explained.

14.6.1 Interplay of AI, blockchain, and IoT

To harness the synergistic effects of AI, blockchain, and IoT in DAM, organizations should identify areas where these technologies complement each other and work in tandem to address key challenges in asset management. For example, AI-driven algorithms can analyze data collected by IoT sensors to derive actionable insights for predictive maintenance or anomaly detection. Blockchain can be used to securely record and verify the provenance and authenticity of assets monitored by IoT devices, while AI algorithms enhance data analysis and decision-making capabilities within blockchain-powered systems. By fostering collaboration and interoperability between AI, blockchain, and IoT components, organizations can create a cohesive ecosystem that leverages the strengths of each technology to optimize asset management workflows and enhance operational efficiency.

14.6.2 Achieving seamless connectivity

Seamless connectivity between AI, blockchain, and IoT devices is essential for ensuring real-time data exchange, interoperability, and responsiveness within the DAM ecosystem. Organizations should invest in robust communication protocols, middleware solutions, and application programming interfaces (APIs) that facilitate seamless integration and interoperability between AI algorithms, blockchain networks, and IoT devices. By standardizing communication protocols and data formats, organizations can overcome compatibility issues and enable seamless connectivity across heterogeneous devices and platforms. Cloud-based infrastructure and edge computing technologies can further enhance connectivity by enabling distributed processing and analysis of data generated by IoT devices, while ensuring scalability and reliability in complex DAM environments.

14.6.3 Cross-platform compatibility

Cross-platform compatibility is critical for ensuring that AI, blockchain, and IoT technologies can seamlessly integrate with existing DAM systems

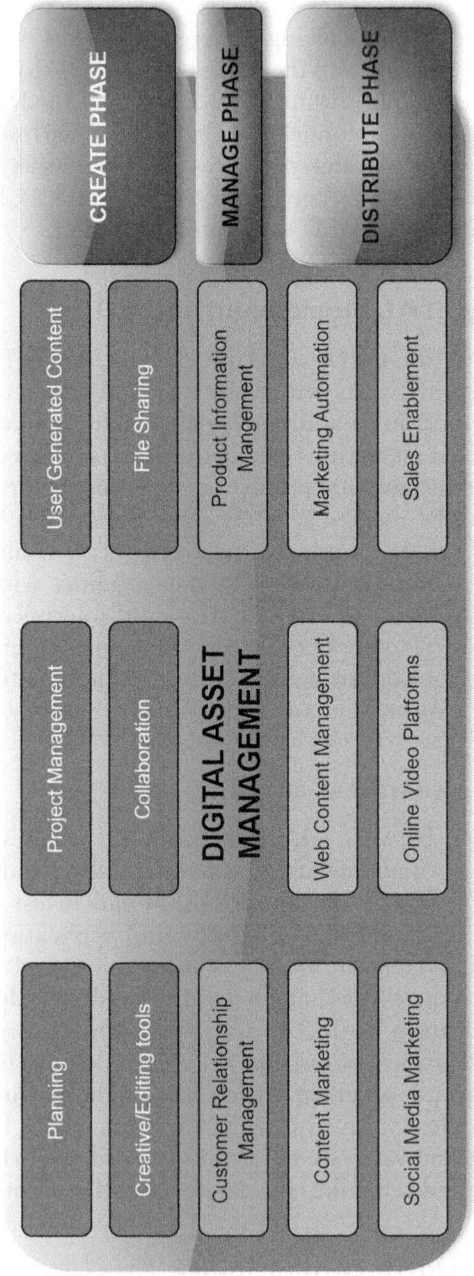

Figure 14.3 Different phases in digital asset management.

and third-party applications. Organizations should prioritize interoperability and compatibility when selecting and deploying technology solutions, ensuring that AI models, blockchain protocols, and IoT devices adhere to industry standards and support open-source frameworks and APIs. Interoperable solutions enable organizations to leverage existing investments in DAM infrastructure while integrating new technologies and functionalities to enhance asset management capabilities. Collaboration with technology partners, participation in industry consortia, and adherence to emerging standards and best practices can help organizations navigate the complexities of cross-platform compatibility and interoperability in the evolving landscape of DAM.

The integration strategies for achieving synergistic effects between AI, blockchain, and IoT in DAM require a holistic approach that emphasizes the interplay between these technologies, seamless connectivity, and cross-platform compatibility. By fostering collaboration, standardization, and interoperability, organizations can unlock new opportunities for innovation and value creation in digital asset management, driving transformative change and competitive advantage in today's digital economy.

14.7 ENHANCING WORKFLOW EFFICIENCY IN DIGITAL ASSET MANAGEMENT

Enhancing workflow efficiency in digital asset management (DAM) involves leveraging technologies and strategies to streamline processes, improve accessibility and searchability, and optimize resource allocation [22].

14.7.1 Streamlining processes with AI

Artificial intelligence (AI) plays a pivotal role in streamlining DAM processes by automating repetitive tasks, accelerating workflows, and enhancing productivity. AI-powered tools can automate metadata tagging, content categorization, and asset organization, reducing manual intervention and streamlining the ingestion and indexing of digital assets. Machine learning algorithms can analyze user behavior and content interactions to personalize search results, recommend relevant assets, and facilitate intuitive navigation within DAM systems. By leveraging AI-driven automation and predictive analytics, organizations can minimize administrative overhead, mitigate human errors, and optimize workflow efficiency across the entire asset lifecycle [23].

14.7.2 Improving accessibility and searchability

Improving accessibility and searchability is essential for ensuring that users can quickly locate, retrieve, and share digital assets within DAM systems.

Advanced search capabilities powered by AI and natural language processing (NLP) enable users to perform complex searches using keywords, filters, and contextual queries, thereby reducing the time and effort required to find relevant assets. AI-driven metadata enrichment enhances the accuracy and richness of asset metadata, improving search precision and discoverability across large asset repositories. User-friendly interfaces, intuitive navigation menus, and customizable dashboards further enhance accessibility and usability, empowering users to seamlessly interact with DAM systems and access assets from any device or platform.

14.7.3 Optimizing resource allocation

Optimizing resource allocation involves aligning organizational resources, including personnel, infrastructure, and technology investments, with strategic priorities and business objectives. AI-driven analytics and performance metrics provide insights into asset usage patterns, content engagement metrics, and resource utilization trends, enabling organizations to optimize resource allocation and investment decisions [24]. By identifying underutilized assets, redundant workflows, and inefficiencies in DAM processes, organizations can reallocate resources, prioritize high-impact initiatives, and maximize the return on investment (ROI) from their DAM infrastructure. Additionally, workflow automation and collaborative tools facilitate seamless coordination and communication among cross-functional teams, optimizing resource allocation and improving collaboration efficiency within DAM ecosystems.

Enhancing workflow efficiency in digital asset management requires a multifaceted approach that integrates AI-driven automation, advanced search capabilities, and optimized resource allocation strategies. By leveraging AI technologies to streamline processes, improve accessibility, and optimize resource allocation, organizations can unlock new levels of productivity, agility, and innovation in managing digital assets effectively. As organizations continue to embrace digital transformation, investing in DAM solutions that prioritize workflow efficiency will be essential for driving sustainable growth and competitive advantage in today's dynamic business landscape.

14.8 SECURITY PROTOCOLS AND BEST PRACTICES

Security protocols and best practices are paramount in digital asset management (DAM) to address privacy concerns, ensure data integrity and confidentiality, and maintain compliance with regulatory standards [25]. Figure 14.4 illustrates the various security features and privacy protocols.

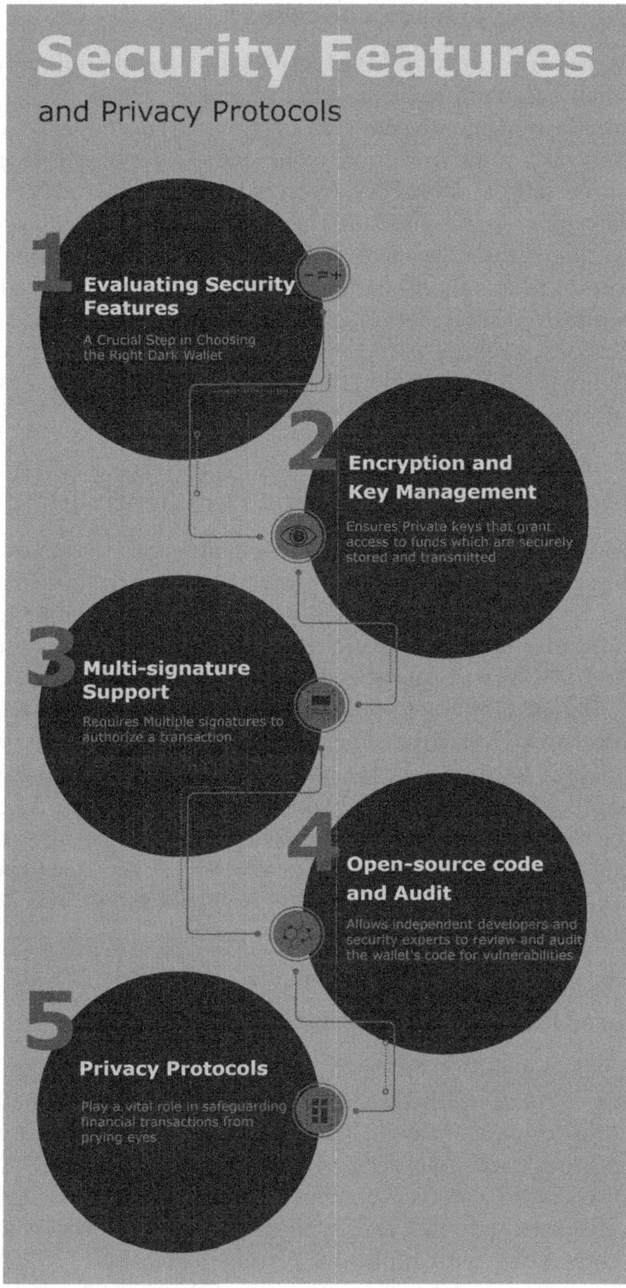

Figure 14.4 Security features and privacy protocols.

14.8.1 Addressing privacy concerns

Privacy concerns are critical in DAM, especially when handling sensitive or personally identifiable information (PII). To address privacy concerns, organizations should implement robust access control mechanisms, encryption protocols, and data anonymization techniques. Role-based access control (RBAC) ensures that users only have access to the assets and functionalities necessary for their roles, minimizing the risk of unauthorized access.

Encryption algorithms, such as AES (Advanced Encryption Standard), should be employed to encrypt data both in transit and at rest, safeguarding it from unauthorized interception or disclosure. Additionally, data anonymization methods, such as pseudonymization and tokenization, can help protect individual privacy while still enabling analysis and utilization of data for legitimate purposes [26].

14.8.2 Ensuring data integrity and confidentiality

Data integrity and confidentiality are fundamental to maintaining the trustworthiness of digital assets and protecting sensitive information from unauthorized access or tampering. Organizations should implement robust authentication mechanisms, strong password policies, and multi-factor authentication (MFA) to verify the identity of users and prevent unauthorized access to DAM systems. Secure communication protocols, such as HTTPS (Hypertext Transfer Protocol Secure), should be used to encrypt data transmission between clients and servers, preventing eavesdropping and data interception. Regular data backups, integrity checks, and audit trails help ensure data integrity by detecting and mitigating unauthorized modifications or data corruption incidents. Moreover, data classification policies and encryption standards should be enforced to protect confidential information and prevent data leakage or exposure.

14.8.3 Compliance with regulatory standards

Compliance with regulatory standards, such as GDPR (General Data Protection Regulation), HIPAA (Health Insurance Portability and Accountability Act), and CCPA (California Consumer Privacy Act), is essential for organizations handling sensitive data and personal information within DAM systems. Organizations should conduct regular privacy impact assessments (PIAs) to evaluate the potential risks and implications of data processing activities and ensure compliance with applicable privacy regulations [27]. Transparent data governance frameworks, data retention policies, and consent management mechanisms should be implemented to maintain compliance with regulatory requirements and facilitate accountability and transparency in data handling practices. Collaboration with legal and compliance teams, as well as regular training and awareness programs for employees, are essential to ensure a proactive approach to compliance and mitigate the risk of regulatory violations.

Addressing privacy concerns, ensuring data integrity and confidentiality, and maintaining compliance with regulatory standards are critical aspects of security protocols and best practices in digital asset management. By implementing robust security measures, organizations can safeguard sensitive information, protect digital assets from unauthorized access or tampering, and build trust with stakeholders, fostering a secure and compliant environment for managing digital assets effectively.

14.9 CASE STUDIES AND PRACTICAL IMPLEMENTATIONS

Case studies and practical implementations offer valuable insights into the real-world applications of integrated technologies in digital asset management (DAM), highlighting success stories, lessons learned, and challenges encountered during implementation.

14.9.1 Real-world applications of integrated technologies

a. **Media and Entertainment Industry**: A media company integrates AI-driven content classification algorithms with blockchain-based digital rights management (DRM) solutions to automate the categorization and secure distribution of digital assets. By leveraging AI for content tagging and classification and blockchain for rights management and royalty tracking, the company enhances efficiency, transparency, and copyright protection in digital asset distribution.
b. **Supply Chain Management**: An industrial manufacturer employs IoT sensor networks, AI analytics, and blockchain technology to optimize supply chain visibility and traceability. IoT sensors embedded in products and packaging capture real-time data on product location, condition, and environmental parameters, while AI algorithms analyze data streams to predict demand, optimize inventory levels, and reduce supply chain inefficiencies. Blockchain ensures the integrity and transparency of supply chain transactions, enabling stakeholders to trace the provenance of products and verify authenticity throughout the supply chain.

14.9.2 Success stories and lessons learned

a. **Financial Services**: A financial institution implements AI-driven anomaly detection systems and blockchain-based audit trails to enhance fraud detection and regulatory compliance in digital asset transactions. By analyzing transactional patterns and identifying anomalous behaviors in real-time, the institution reduces false positives and improves fraud detection accuracy. Lessons learned include the importance of data quality, model interpretability, and continuous monitoring to adapt to evolving threats and regulatory requirements.

b. **Healthcare Sector**: A healthcare organization integrates IoT-enabled medical devices, AI-powered predictive analytics, and blockchain-based patient records management systems to improve patient outcomes and data security. IoT devices collect patient health data in real-time, while AI algorithms analyze data trends to predict disease progression and recommend personalized treatment plans. Blockchain ensures the integrity and confidentiality of patient records, facilitating secure data sharing and interoperability among healthcare providers.

14.9.3 Challenges and solutions in implementing emerging technologies

a. **Data Privacy and Security**: Implementing AI, Blockchain, and IoT technologies requires robust data privacy and security measures to protect sensitive information and prevent unauthorized access. Solutions include implementing encryption protocols, access controls, and data anonymization techniques to safeguard data confidentiality and integrity.
b. **Interoperability and Integration**: Integrating disparate technologies and systems poses challenges related to interoperability, data exchange formats, and integration complexity. Solutions involve adopting open standards, APIs, and middleware solutions to facilitate seamless communication and interoperability between different technologies and platforms.

Case studies and practical implementations demonstrate the diverse applications of integrated technologies in DAM, highlighting success stories, lessons learned, and challenges encountered. By understanding real-world use cases and best practices, organizations can leverage emerging technologies to enhance efficiency, transparency, and security in digital asset management, driving innovation and value creation across various industries and domains.

14.10 CONCLUSION

The integration of emerging technologies such as artificial intelligence (AI), blockchain, and the Internet of Things (IoT) holds immense potential for transforming digital asset management (DAM) practices, enhancing security, efficiency, and transparency across diverse industries and domains. Through the interplay of AI, blockchain, and IoT, organizations can streamline processes, automate workflows, and optimize resource allocation, thereby unlocking new opportunities for innovation and value creation in DAM ecosystems. AI-driven algorithms facilitate automated content classification, metadata enrichment, and anomaly detection, empowering organizations to manage digital assets intelligently and proactively. Blockchain technology offers a transparent, tamper-resistant ledger for recording asset transactions, ensuring data integrity, and enabling verifiable provenance and authenticity verification. By leveraging blockchain's decentralized architecture, organizations can establish secure, immutable records of digital assets, fostering trust and transparency in asset management workflows. The Internet of Things (IoT) enables real-time asset monitoring, leveraging asset tracking, sensor networks, and predictive analytics to optimize asset utilization, enhance operational efficiency, and mitigate risks associated with asset downtime and failure. IoT devices embedded within assets provide granular insights into asset performance, condition, and environmental parameters, enabling organizations to implement predictive maintenance strategies and proactive asset management practices. Success stories and practical implementations demonstrate the tangible benefits of integrating AI, blockchain, and IoT in DAM, including improved workflow efficiency, enhanced security, and compliance with regulatory standards. However, challenges such as data privacy concerns, interoperability issues, and integration complexities underscore the importance of adopting a holistic approach to technology adoption and implementation. In navigating the evolving landscape of DAM, organizations must prioritize collaboration, innovation, and continuous improvement, leveraging emerging technologies to address evolving business needs and stakeholder expectations. By embracing the transformative potential of AI, blockchain, and IoT, organizations can fortify their digital asset security, optimize operational efficiency, and drive sustainable growth in today's digital economy. Thus, the integration of AI, blockchain, and IoT technologies offers a roadmap for organizations seeking to navigate the complexities of digital asset management successfully. By synthesizing these emerging technologies, organizations can unlock new levels of efficiency, transparency, and value creation, positioning themselves for success in an increasingly digitized world.

REFERENCES

1. Humphrey, C. D., Tollefson, T. T., & Kriet, J. D. (2010). Digital asset management. *Facial Plastic Surgery Clinics*, 18(2), 335–340.
2. McCord, A. (2002). Overview of digital asset management systems. EDUCAUSE Evolving Technologies Committee. *Applied Sciences*. 12(15):7694.
3. Keathley, E. F., De Gyor, H., & Windsor, R. (2014). *Digital asset management: content architectures, project management, and creating order out of media chaos*. Apress.

4. Re Cecconi, F., Dejaco, M. C., Moretti, N., Mannino, A., & Blanco Cadena, J. D. (2020). *Digital asset management. Digital transformation of the design, construction and management processes of the built environment* (pp. 243–253).
5. Cherrington, M., Lu, Z., Xu, Q., Thabtah, F., Airehrour, D., & Madanian, S. (2020). Digital asset management: New opportunities from high dimensional data—A New Zealand perspective. In *Advances in asset management and condition monitoring: Comadem 2019* (pp. 183–193). Springer International Publishing.
6. Chimakurthi, V. N. S. S. (2020). Digital asset management in the communication of product promotional activities. *Asian Business Review*, 10(3), 177–186.
7. McGovern, M. (2013). Digital asset management: where to start. *Curator: The Museum Journal*, 56(2), 237–254.
8. Raja, G. B. (2023). Cloud-based remote sensing: Developments and challenges—research point of view. In *Big data, cloud computing and IoT* (pp. 65–80). Chapman and Hall/CRC.
9. Raja, G. B. (2021). Impact of internet of things, artificial intelligence, and blockchain technology in Industry 4.0. In *Internet of things, artificial intelligence and blockchain technology* (pp. 157–178).
10. Raja, G. B. (2021). Impact of IIOT in future industries: Opportunities and challenges. *Internet of Things*, 21, 101–112.
11. Novick, B., Mayston, D., Marcus, S., Barry, R., Fox, G., Betts, B., ... & Eisenmann, K. (2019, Oct). *Artificial intelligence and machine learning in asset management*. Blackrock.
12. Mattioli, J., Perico, P., & Robic, P. O. (2020, June). Artificial intelligence based asset management. In *2020 IEEE 15th International Conference of System of Systems Engineering (SoSE)* (pp. 151–156). IEEE.
13. Cherrington, M., Lu, Z., Xu, Q., Airehrour, D., Madanian, S., & Dyrkacz, A. (2020). Deep learning decision support for sustainable asset management. In *Advances in asset management and condition monitoring: COMADEM 2019* (pp. 537–547). Cham: Springer International Publishing.
14. Li, J., Herdem, M. S., Nathwani, J., & Wen, J. Z. (2023). Methods and applications for artificial intelligence, big data, Internet of Things, and blockchain in smart energy management. *Energy and AI*, 11, 100208.
15. Lemieux, V., Hofman, D., Batista, D., & Joo, A. (2019). *Blockchain technology & recordkeeping*. ARMA International Educational Foundation.
16. Biswas, S., Sharif, K., Li, F., Alam, I., & Mohanty, S. P. (2020). DAAC: Digital asset access control in a unified blockchain based e-health system. *IEEE Transactions on Big Data*, 8(5), 1273–1287.
17. Zhu, Y., Qin, Y., Zhou, Z., Song, X., Liu, G., & Chu, W. C. C. (2018, July). Digital asset management with distributed permission over blockchain and attribute-based access control. In *2018 IEEE International Conference on Services Computing (SCC)* (pp. 193–200). IEEE.
18. Sarkar, D., Patel, H., & Dave, B. (2022). Development of integrated cloud-based Internet of Things (IoT) platform for asset management of elevated metro rail projects. *International Journal of Construction Management*, 22(10), 1993–2002.
19. Deese, A. S., Jesson, J., Brennan, T., Hollain, S., Stefanacci, P., Driscoll, E., ... & Rodriguez, E. (2020). Long-term monitoring of smart city assets via Internet of Things and low-power wide-area networks. *IEEE Internet of Things Journal*, 8(1), 222–231.

20. Feizabadi, J., Gligor, D. M., & Alibakhshi, S. (2021). Examining the synergistic effect of supply chain agility, adaptability and alignment: A complementarity perspective. *Supply Chain Management: An International Journal*, 26(4), 514–531.
21. Chen, J. L. (2012). The synergistic effects of IT-enabled resources on organizational capabilities and firm performance. *Information & Management*, 49(3–4), 142–150.
22. Love, P. E., & Matthews, J. (2019). The 'how' of benefits management for digital technology: From engineering to asset management. *Automation in Construction*, 107, 102930.
23. Rana, S. K., & Rana, S. K. (2021). Intelligent amalgamation of blockchain technology with industry 4.0 to improve security. In *Internet of Things* (pp. 165–175). CRC Press.
24. Jiang, J. N., Henning, T. F. P., & Zou, Y. (2022). Digital transformation in asset management–A case of BIM adoption in New Zealand local government. In *ISARC. Proceedings of the International Symposium on Automation and Robotics in Construction* (Vol. 39, pp. 574–581). IAARC Publications.
25. Usmani, U. A., Happonen, A., & Watada, J. (2023, October). Advancements in industry 4.0 asset management: Interoperability and cyber security challenges and opportunities. In *Proceedings of the future technologies conference* (pp. 468–488). Cham: Springer Nature Switzerland.
26. Chimakurthi, V. N. S. S. (2020). Digital asset management: A lowdown on intricacies of digital rights and permissions. *Global Disclosure of Economics and Business*, 9(2), 129–140.
27. Kuhle, P., Arroyo, D., & Schuster, E. (2021). Building a blockchain-based decentralized digital asset management system for commercial aircraft leasing. *Computers in Industry*, 126, 103393.

Chapter 15

Secure and intelligent data-driven technique with IPv6 in Internet of Things applications

Arun Kumar Rana, Sumit Kumar Rana, Ritu Dewan, Amit Kumar Kesarwani, Ajay Beniwal and Shikha Singh

15.1 INTRODUCTION

It is expected that the IoT will have profound effects on people's daily lives. The analysis demonstrates the dramatic increase in the number of devices owned by each person, with an average of over eight by 2020. As a result, the cost of the devices dropped, making them more affordable for regular people. Observation layer, organizing layer, administration layer, and application layer make up the IoT's multilayer architecture [1]. Each device in the Internet of Things (IoT) has a unique location determined by the universally accepted addressing standards. 'Internet' refers to the global network of interconnected personal computer systems based on the (TCP/IP) correspondence rules, while 'Thing' refers to any device or gadget used in the Internet of Things. With the help of IoT, users may maintain a constant connection to their far-flung devices. Any method, structure, or aid can be used to establish a connection between electronic devices. If the correspondence mechanism communicates the data between the heterogeneous devices via heterogeneous systems, then several types of correspondence can be used in IoT. While IoT's device-to-device communication necessitates its own unique guiding conventions, there are many scenarios in which we may benefit from having more flexible guiding protocols. One of the established norms for navigating IoT networks is RPL [2]. Investigate the RPL standard now by checking its protection against a variety of IoT-based attacks. Contributions of paper are

- It discusses the latest papers investigating the Security.
- It proposes a secure Caesar Cypher hash algorithm, which is used to enhance the security in IoT application.
- It proposes solutions using IPv6 to improves security when compared with an existing algorithm Caesar Cypher hash algorithm

 DOI: 10.1201/9781003590439-15

15.2 LITERATURE REVIEW

To help customers better comprehend and adapt to service disruptions, Jie Lin et al. [1] suggested a fog-based framework for the Internet of Things to enable pro-processing services gadgets sent at the system edge. Due to the advantages of distributed computing and proximity to end users, mist/ edge computing can improve the responsiveness and quality of service provided by Internet of Things (IoT) applications. Atmel addressed concerns regarding IoT security in a white paper [2]. Basic system issues that arise before the alert reaches the edge hub were discussed, as were recent developments in web security like SSL and TLS. Those hubs whose credibility has been verified can now take use of the perks that have been established for them. This work has identified several types of pull in at the hub of the edge (see [10–12]). Finally, by employing crypto quickening agent devices, they have provided a perceptive response for encryption at the edge hub. For IoT smart city deployments, Container Cheng et al. [3] have presented a standard-based approach to managing the planning and implementation of another mist figure based system, namely Fog Flow. They have also discussed how to improve task allocation for low idle times and transmission capacity utilization, as well as how to more effectively organize and manage information preparation projects across cloud and edges.

Lightweight validation and verified steering (LASeR) for NDN (Named Data Networking IoT in Smart Cities [4], a secure onboarding and directing system for) NDN-based IoT arrangement is proposed by Travis Mick, Reza Tourani, and Satyajayant Misra. This is now achieved via an evolving system architecture with negligible cryptographic or computing overhead. Federico Montoriononi offered [5] an extensible engineering that nonetheless allows for the coexistence of different methods of information collection. A solution based on human use of context and judgement is proposed [6] by Joshua E. Siegel et al., who advocate using cloud resources to promote IoT on mandated devices. They provide security via reflection and protection via distant information combination by presenting engineering application process information. The Data Proxy enables low-resource analysis of a framework's sensors in accordance with a specified Nature of Data focus. After being implemented, this design's enhanced productivity was first seen in a prototype car. Finally, they consider potential future avenues for this architecture to lessen technical, financial, and conclusion barriers to the use of the IoT.

A secure and effective information gathering strategy for Cloud-IoT in the smart network was proposed by Zhitao Guan et al. [7]. The proposed method involves partitioning the massive amount of data into smaller squares, which are then encoded/decoded and transmitted together. XILINX's white paper [8] examines the development of an IoT network, cyber security, and edge figure within the context of selecting an IoT edge stage that can adapt to the influence of market drifts over time. In this study,

José V. V. Sobral et al. [9] examine the latest methods utilizing IoT and RPL. This study will focus on expanding existing RPL routing protocol-based security situations. In addition, they are given the double-mystery-sharing-plan [10], which takes into account the checking of attributes with the checking of informational honesty and security. The research demonstrates that the proposed plan may satisfy the security requirements of information acquisition in the dazzling framework, while also reducing the reaction time overhead totally in comparison to other prevalent strategies. Different problems and possible solutions, including increased power, storage, and bandwidth, are discussed in this work by Sharwari S. Solapure et al. [11].

15.3 RESULT AND ANALYSIS

IoT A computer operating system, Contiki is comparable to Windows and Linux. However, it zeroed down on IoT-related "matters" for the sake of a more precise explanation. Programming/system control, valuable asset management, executive memory, and board-level verbal exchange are all parts of an operating system. With smart soil and a focus on small devices, Contiki OS aims to meet all of your needs. Things should be able to exchange limited amounts of information with one another. The Contiki OS is an Internet of Things operating system made to aid in organized, asset-driven devices. With just 2 kilobytes of RAM and 60 kilobytes of ROM running at 1 MHz [7], a typical setup can be conveyed using the C-implemented Contiki's lightweight memory the board and force efficiency. In 2017, Contiki was used in about 13% of devices, and that number is only expected to rise, as reported by the IoT engineer review [12–18]. With the goal of connecting small, battery-operated devices to the internet, Contiki offers lightweight utilization of a number of popular communication principles as IEEE 802.15.4, 6LoWPAN, CoAP, MQTT, TSCH, and RPL. In addition, Contiki features an independent programming framework for equipment, with the core framework providing a reasonable level of reflection. This promotes framework flexibility, as additional stage support may be implemented in libraries and

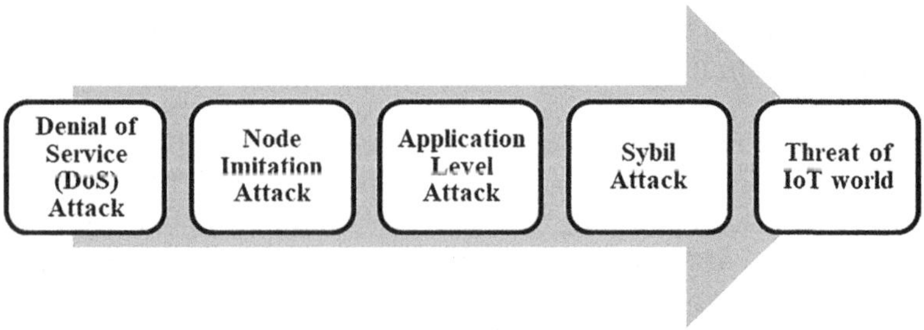

Figure 15.1 Types of attack in IoT network.

services on top of Contiki's liquid design, which is especially useful given the indisputably application-driven nature of sensor devices [19–22].

Types of Attack in IoT network is shown in Figure 15.1. Technology has progressed to the point that machines no longer require human assistance to communicate with one another. In addition, we call this network of connections the web of things. Among the many things that rely on IoT communication are smart cities, commercial hubs, barrier hardware, traffic systems, highway monitoring, smart workplaces, dazzling cost collection, satellite television, and smart homes [23–27]. These devices are effectively interconnected, and communication takes place mechanically. Using the Contiki cooja IPv6 standards, the system is extremely safe. By disrupting the devices' ability to communicate and exchange data, malicious packages or the aggressor hub complete a denial-of-service attack. With a fake ID, an attacker can impersonate a hub and take control of it. Consequently, the hub will get the irrelevant communications. The attacker can then use this as a springboard to attack the entire IOT infrastructure. With an Application-Level Attack, the aggressor first attacks the hub, then modifies the message before sending it back to the target hub. The attacker can send false information to the hubs via this method of attack. In a Sybil Attack, the attacker creates a copy of the source hub, assumes control of the personality, and uses the power of other hubs to flee the system. Figure 15.2 Shows features introduced with the IPv6 protocol for RPL.

The main problem with IoT devices is information security and data integrity. There must be encrypted communication between the devices. IPv6's convention ensures very secure communication. Caesar Cipher's dynamic cryptographic confirmation and key age will make IPv6-based computations extremely safe. RPL is an IPv6-based standard that is commonly used in the Internet of Things. And it's mostly coordinated with

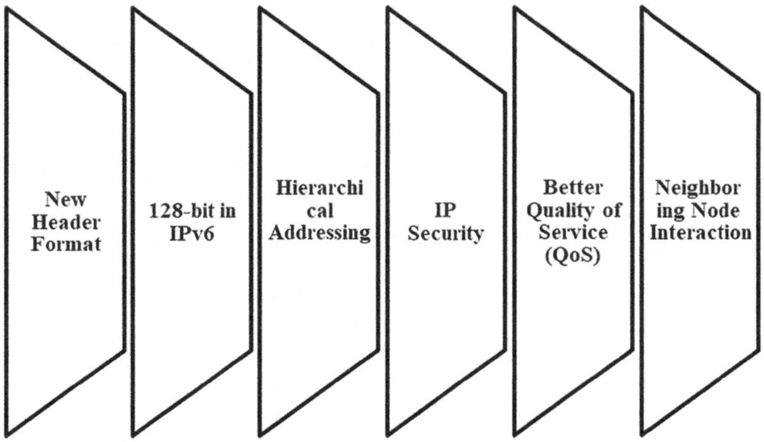

Figure 15.2 Features introduced with the IPv6 protocol for RPL.

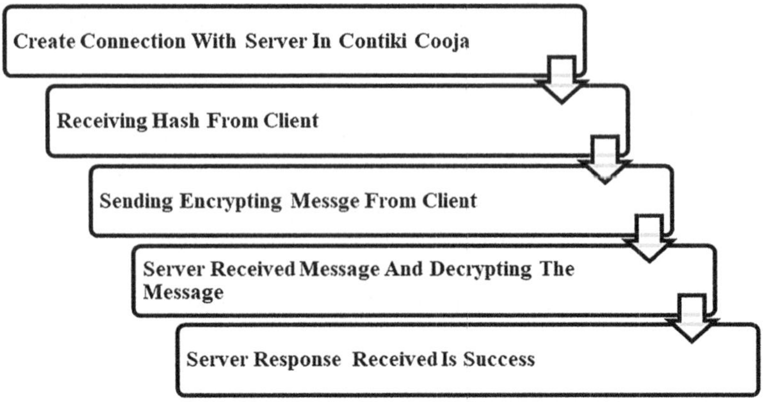

Figure 15.3 Security solution with IPv6 RPL routing.

systems in low-force, out-of-the-way areas. Figure 15.3 is a flowchart depicting the IPv6 RPL routing security solution.

Simulation scenarios with 1 sink (green color) with 6 senders (yellow color) with RPL routing. is shown in Figure 15.4. Authentication between two nodes is a necessary first step in any successful communication. The hash is computed on the client side and then transmitted to the server. If the server's calculated hash from previously stored data matches the sender's, I'll send an acknowledgment. In order to encrypt a message, the sender must first get an acknowledgment from the recipient, after which the server will transmit the client the key's coordinates. If the server receives a key whose coordinates match those saved in the client's database, the message is decrypted successfully; otherwise, the client is disconnected. Figures 15.5

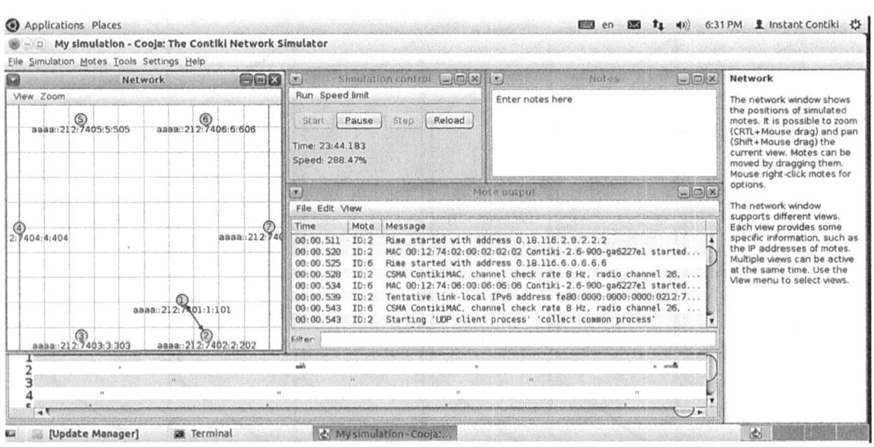

Figure 15.4 Simulation scenarios with 1 sink (green color) with 6 senders (yellow color) with RPL routing.

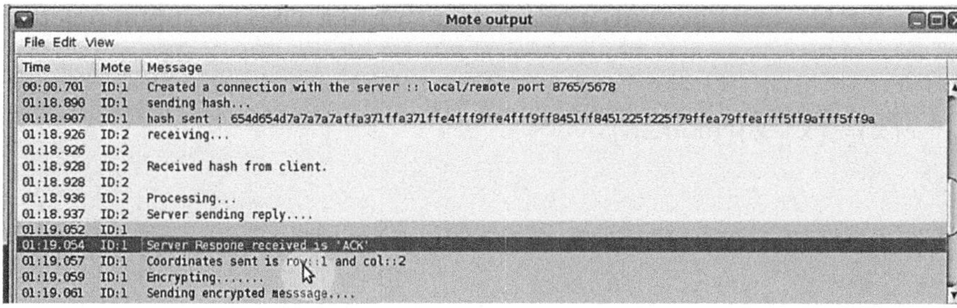

Figure 15.5 Encryption process.

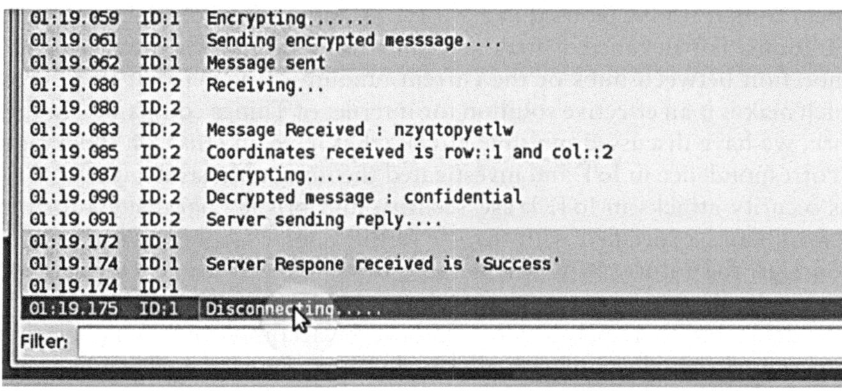

Figure 15.6 Decryption process.

and 15.6 depict the encryption and decryption procedures used to ensure the confidentiality of sensitive information. IPsec security is planned in the IPv6 convention detail, allowing IPv6 package validation and maybe payload encryption via Extension Headers, which was not the case with IPv4. IPsec, however, isn't always active; typically, it requires configuration and use in conjunction with a security key exchange.

In our **mathematical model** First translate all of our characters to numbers, 'a' = 0, 'b' = 1, 'c' =2 … 'z' = 25 and demonstrate the Caesar cipher encryption function, e(x), where x is the character we are encrypting, as shown in Figure 15.7:

$$e(x) = (x + k)(\text{mod } 26)$$

Figure 15.7 Caesar cipher mathematical model.

where k is the key (the shift) applied to each letter. The decryption function is:

$$e(x) = (x - k)(\text{mod } 26)$$

15.4 CONCLUSIONS AND FUTURE SCOPE

The Internet of Things (IoT) network is a restricted network in terms of energy, computing power, memory, and so on; as a result, it is extremely vulnerable to security assaults. RPL also makes it possible to develop the system so that it may be used in a variety of application scenarios and configurations. For instance, it may take into consideration the quality of the connection between hubs or the current amount of power that they have, which makes it an effective solution for Internet of Things companies. In this work, we have discussed multiple attacks that have an effect on the nature of correspondence in IoT and investigated the mitigation strategies for various security attacks in IoT. Trust, security, and privacy parameters for the network can be specified with the use of the Caesar Cypher Mathematical Model. In the future, RPL will be able to be utilized for 5G technology, which will allow 5G to provide an industrial IoT network that is both more dependable and extremely secure. This will be accomplished by integrating RPL security into the core network design.

REFERENCES

[1] Lin Jie, Wei Yu, Nan Zhang, Xinyu Yang, Hanlin Zhang. A survey on internet of things: Architecture enabling technologies security and privacy and applications. *IEEE Internet of Things Journal*, Volume: 4, pp. 2327–4662, 2015.

[2] Atmel-8994-Security-for-Intelligent-Connected-IoT-Edge-Nodes. Atmel Corporation. Rev.:Atmel-8994A-CryptoAuth-Security-for-Intelligent-Connected-IoT-Edge NodesWhitePaper_112015, 2015.

[3] Cheng Bin, Solmaz, Gürkan, Cirillo, Flavio, Kovacs, Ernö, Terasawa, Kazuyuki, and Kitazawa, Atsushi. FogFlow: Easy Programming of IoT Services Over Cloud and Edges for Smart Cities. *IEEE Internet of Things Journal*, Volume: 2, pp. 2327–4662, 2017.

[4] Travis, Mick, Tourani, Reza, and Misra, Satyajayant. LASeR: Lightweight Authentication and Secured Routing for NDN IoT in Smart Cities. *IEEE Internet of Things Journal*, pp. 2327–4662, 2017.

[5] Montori, Federico, Bedogni, Luca, and Bononi, Luciano. A Collaborative Internet of Things Architecture for Smart Cities and Environmental Monitoring. *IEEE Internet of Things Journal*, Volume: 4, pp. 2327–4662, 2017.

[6] Siegel, Joshua E, Kumar, Sumeet, and Sarma, Sanjay E.. The Future Internet of Things: Secure, Efficient, and Model-Based. *IEEE Internet of Things Journal*, pp. 2327–4662, 2017.

[7] Guan, Zhitao, Jing Li, Longfei Wu, Yue Zhang, Jun Wu, and Xiaojiang Du. Achieving Efficient and Secure Data Acquisition for Cloud-Supported Internet of Things in Smart Grid *IEEE Internet of Things Journal*, Volume: 4, pp. 2327–4662, 2017.

[8] Khona, Chetan. Key Attributes of an Intelligent IIoT Edge Platform. White Paper: All Programmable Devices XILINX. WP493 (v1.0) September 6, 2017.

[9] José, V. V. Sobral, Routing Protocols for Low Power and Lossy Networks in Internet of Things Applications Sensors 2019.

[10] Krishna, Rama, Dhawan, Sachin, Sharad Sharma and Arunkumar Review on Artificial Intelligence with Internet of Things - Problems, Challenges and Opportunities, In *2nd International Conference on Power Energy, Environment and Intelligent Control (PEEIC)*, Greater Noida, India, pp. 383–387. 2019

[11] Solapure, S.S., Kenchannavar, H.H., and Sarode, K.P. Issues Faced During RPL Protocol Analysis in Contiki–2.7. In: M. Tuba, S. Akashe, A. Joshi (Eds.), *ICT Systems and Sustainability. Advances in Intelligent Systems and Computing*, vol 1077. Springer, Singapore, 2020.

[12] Rana, A., Sharma, S., Nisar, K., Ibrahim, A., Ag, A., Dhawan, S., ... and Goyal, N. (2022). The Rise of Blockchain Internet of Things (BIoT): Secured, Device-to-Device Architecture and Simulation Scenarios. *Applied Sciences*, 12(15), 7694.

[13] Kumar, A., Sharma, S., Goyal, N., Singh, A., Cheng, X., and Singh, P. (2021). Secure and Energy-Efficient Smart Building Architecture With Emerging Technology IoT. *Computer Communications*, 176, 207–217.

[14] Kumar, A., Sharma, S., Singh, A., Alwadain, A., Choi, B. J., Manual-Brenosa, J., ... and Goyal, N. (2021). Revolutionary Strategies Analysis and Proposed System for Future Infrastructure in Internet of Things. *Sustainability*, 14(1), 71.

[15] Rana, A., Chakraborty, C., Sharma, S., Dhawan, S., Pani, S. K., and Ashraf, I. (2022). Internet of Medical Things-Based Secure and Energy-Efficient Framework for Health Care. *Big Data*, 10(1), 18–33.

[16] Kumar, A., and Sharma, S. (2021). Internet of Robotic Things: Design and Develop the Quality of Service Framework for the Healthcare Sector Using CoAP. *IAES International Journal of Robotics and Automation*, 10(4), 289.

[17] Dhawan, S., Chakraborty, C., Frnda, J., Gupta, R., Rana, A. K., and Pani, S. K. (2021). SSII: Secured and High-Quality Steganography Using Intelligent Hybrid Optimization Algorithms for IoT. *IEEE Access*, 9, 87563–87578.

[18] Rana, S. K., Kim, H. C., Pani, S. K., Rana, S. K., Joo, M. I., Rana, A. K., and Aich, S. (2021). Blockchain-Based Model to Improve the Performance of the Next-Generation Digital Supply Chain. *Sustainability*, 13(18), 10008.

[19] Rana, A. K., and Sharma, S. (2021). Internet of Things Based Stable Increased-throughput Multi-hop Protocol for Link Efficiency (IoT-SIMPLE) For Health Monitoring Using Wireless Body Area Networks. *International Journal of Sensors Wireless Communications and Control*, 11(7), 789–798.

[20] Kumar, A., Sharma, S., Goyal, N., Gupta, S. K., Kumari, S., and Kumar, S. (2022). Energy-Efficient Fog Computing in Internet of Things Based on Routing Protocol for Low-Power and Lossy Network With Contiki. *International Journal of Communication Systems*, 35(4), e5049.

[21] Pandit, M., Gupta, D., Anand, D., Goyal, N., Aljahdali, H. M., Mansilla, A. O., ... and Kumar, A. (2022). Towards Design and Feasibility Analysis of DePaaS: AI Based Global Unified Software Defect Prediction Framework. *Applied Sciences*, 12(1), 493.

[22] Lilhore, U. K., Imoize, A. L., Lee, C. C., Simaiya, S., Pani, S. K., Goyal, N., ... and Li, C. T. (2022). Enhanced Convolutional Neural Network Model for Cassava Leaf Disease Identification and Classification. *Mathematics*, 10(4), 580.

[23] Rana, S. K., Rana, S. K., Nisar, K., Ag Ibrahim, A. A., Rana, A. K., Goyal, N., and Chawla, P. (2022). Blockchain Technology and Artificial Intelligence Based Decentralized Access Control Model to Enable Secure Interoperability for Healthcare. *Sustainability*, 14(15), 9471.

[24] Rana, A. K., and Sharma, S. (2019). Enhanced Energy-Efficient Heterogeneous Routing Protocols in WSNs for IoT Application. *IJEAT*, 9(1), 4415–4418.

[25] Rana, A. K., and Sharma, S. (2021). Industry 4.0 Manufacturing Based on IoT, Cloud Computing, and Big Data: Manufacturing Purpose Scenario. In *Advances in Communication and Computational Technology* (pp. 1109–1119). Springer, Singapore.

[26] Rana, A. K., Krishna, R., Dhwan, S., Sharma, S., and Gupta, R. (2019, October). Review on Artificial Intelligence With Internet of Things-Problems, Challenges and Opportunities. In *2019 2nd International Conference on Power Energy, Environment and Intelligent Control (PEEIC)* (pp. 383–387). IEEE.

[27] Rana, A. K., and Sharma, S. (2021). Contiki Cooja Security Solution (CCSS) With IPv6 Routing Protocol for Low-Power and Lossy Networks (RPL) in Internet of Things Applications. In *Mobile Radio Communications and 5G Networks* (pp. 251–259). Springer, Singapore.

Index

For Product Safety Concerns and Information please contact our EU representative GPSR@taylorandfrancis.com Taylor & Francis Verlag GmbH, Kaufingerstraße 24, 80331 München, Germany

Printed and bound by CPI Group (UK) Ltd, Croydon, CR0 4YY
14/09/2026
02216223-0019